Economical Bridge Solutions based on innovative composite dowels and integrated abutments

Edward Petzek · Radu Băncilă (Eds.)

Economical Bridge Solutions based on innovative composite dowels and integrated abutments

Ecobridge

Editors
Prof. Edward Petzek
Prof. Radu Băncilă
Timişoara, Romania

ISBN 978-3-658-14088-5
DOI 10.1007/978-3-658-06417-4

ISBN 978-3-658-06417-4 (eBook)
ISBN 978-3-658-06418-1 (MyCopy)

The Deutsche Nationalbibliothek lists this publication in the Deutsche Nationalbibliografie;
detailed bibliographic data are available in the Internet at http://dnb.d-nb.de.

Springer Vieweg
© Springer Fachmedien Wiesbaden 2015
Softcover reprint of the hardcover 1st edition 2015

Printed on acid-free paper

Springer Vieweg is a brand of Springer DE. Springer DE is part of Springer Science+Business Media
www.springer-vieweg.de

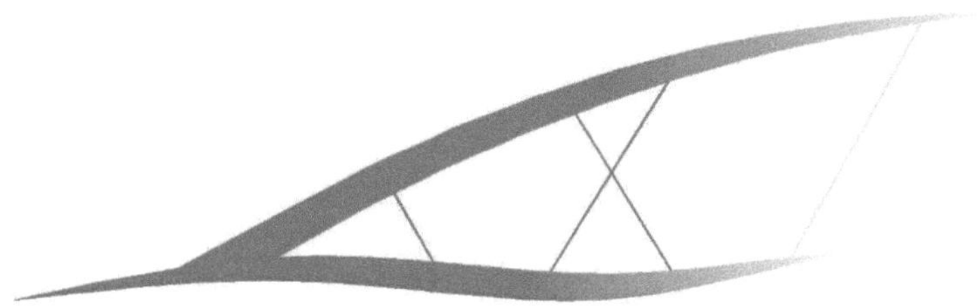

Foreword

This book is an outcome of the research project "**ECOBRIDGE** – *Demonstration of ECO-nomical **BRIDGE** solutions based on innovative composite dowels and integrated abutments* – RFCS – CT 2010-00024", which has been co-funded by the Research Fund for Coal and Steel (R.F.C.S.) of the European Community.

The design and construction of sustainable and durable bridges with low maintenance costs is one of the European Road and Railway Administration tasks. The structures must be safe, economical and with good serviceability issues. All these needs can be found in integral abutment bridges. This solution, by eliminating the bearing and expansion joints, leads to low production and maintenance costs. Integral bridges have a good earthquake resistance. Nevertheless, the design and construction of these structures include a series of specific aspects.

The knowledge gained within the RFCS research projects, INTAB and PRECOBEAM, has enabled us to elaborate cost effective, environmental friendly and sustainable bridge structures. The main objective of the present project entitled "EcoBridge" was the construction of three composite bridges with integral abutments and innovative form of shear transmission – composite dowels. The project focuses on cost efficient, competitive composite bridges with special regard to environmental friendly and sustainable design. The targeted countries are: Germany, Romania and Poland. The bridges were instrumented with a variety of strain gauges, displacement sensors and thermocouples to monitor and help in the assessment of structural behaviour, for future application of integral abutment bridges and/or composite dowels. The consortium members (involving three universities, three design offices, three steel contractors and one steel manufacturer) have long track record and extensive experience in this field. Additionally, the exchange of technical experience from the partners from different parts of Europe contributes to a transfer of information from the different construction fields with different demands.

The main topics of the book are the following: design of integral bridges, innovative composite dowels for the shear transmission, construction of bridges, structural analysis of bridges and monitoring. The book joins the technical experience and the contributions of the involved research partners. The technical content of all the papers is present-day in the field of the design, construction and monitoring of innovative composite bridges. The efficient design and construction improve and consolidate the market position of steel construction and steel producing industry. In addition, the advanced forms of construction are contributing to savings in material and energy consumption for the structure during production and maintenance. All time savings in construction and maintenance result in large benefits for the bridge owners but also for the community as less disturbance of the traffic will occur.

In the frame of the "Eight Danube Bridges Conference" held in Timişoara on the 4th and 5th of October 2013, a special workshop dedicated to the EcoBridge research project was organised (Fig. 1).

Figure 1:   The Eight International Conference on Bridges across the Danube,

"BRIDGES IN DANUBE BASIN – New trends in bridge engineering and efficient solutions for large and medium span bridges"

The International Conference on Bridges across the Danube has become a traditional international event in bridge engineering. Starting with 1992, it is organized periodically each third year in another Danube country.

The First Conference on Danube Bridges – initiated by Prof. Miklos Ivanyi – was held in 1992 on a ship, sailing on the Danube from Vienna via Bratislava to Budapest. The Second Conference was organized in 1995 in Bucharest, the Third Conference was held in 1998 in Regensburg, the Fourth Conference in 2001 in Bratislava, the Fifth Conference took place in Novi Sad in 2004, the Sixth Conference was in Budapest in 2007 and the Seventh Conference in Sofia in 2010.

The Eight International Conference on Bridges across the Danube, titled: "BRIDGES IN DANUBE BASIN – New trends in bridge engineering and efficient solutions for large and medium span bridges" took place in Timişoara (Romania) and in Serbia on the 4th and 5th of October 2013. During the two working days of the Conference, plenary lectures held by invited speakers and selected papers were presented. On October 4th a special meeting dedicated to the European ECOBRIDGE project was organized. A technical site visit in Serbia (Novisad, Belgrade, Zemun) on October 5th presenting some new bridges across the Danube, took place.

The general aim of the Conference was the overall exchange of knowledge and experience between different institutions, owners, contractors, bridge designers and constructors as well as scientific experts. The conference intends also to promote advances in bridge engineering and understanding between the countries along the Danube, but also between other European countries.

The Conference was under the aegis of important scientific organizations: International Association for Bridges on the Danube (IABD), International Association for Bridge and Structural Engineering (IABSE), European Convention for Constructional Steelwork, Universitatea "Politehnica" Timişoara and the Romanian Academy for Technical Sciences (ASTR).

The proceedings of the Conference were published in the Springer – Vieweg Verlag; the volume comprises 42 papers on 544 pages. The attendance was very good (over 90 persons).

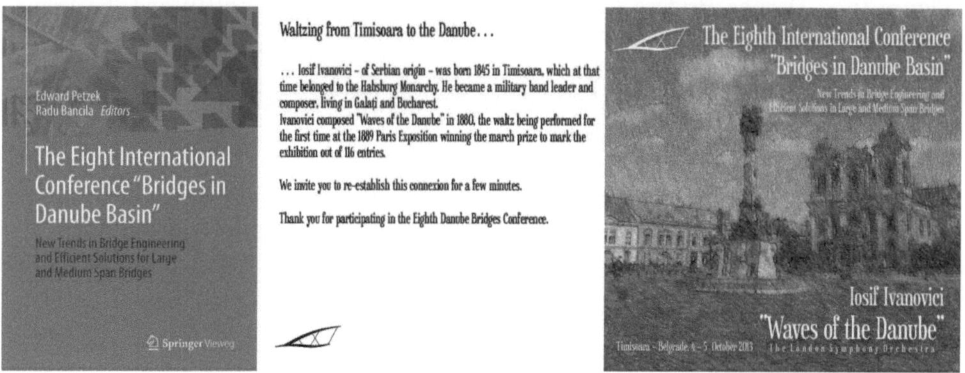

Figure 2: The proceedings of the conference

On the laudable initiative of Prof. Miklos Ivanyi the proceedings of all the Danube Bridges Conferences were recorded on a special CD. It is important to mention that on the CD there is also a reproduction of the last version of the Danube Bridges Catalogue (Edition 1997 Regensburg).

Figure 3:   Danube Bridges Conferences Proceedings (1992 – 2010) and the Danube Bridges Catalogue

The next Danube Bridges Conference will be organized by Slovakia in 2016.
Finally, we want to express our gratitude to:

- the Research Fund for Coal and Steel (RFCS) of the European Community;
- our collaborators – project partners, which contributed to the present publication;
- to the non-profit association SALVPOD, who co-financed – together with RFCS this book.

In memoriam   † *Prof. Dr. Ing. Miklos IVÁNYI*
                          † *Prof. Dr. Ing. Dragoş TEODORESCU*

Timişoara, February 2014
Edward Petzek   **Radu Băncilă**

Acknowledgement for publication

This project has been funded with support from the European Commission. This publication reflects the views only of the authors, and the Commission cannot be held responsible for any use which may be made of the information contained therein.

ECOBRIDGE – ECOnomical BRIDGE solutions based on
innovative composite dowels and integrated abutments
Grant Agreement Number: RFSP-CT-2010-00024

Contents

Demonstration of ECOnomical BRIDGE solutions based on innovative composite dowels and integrated abutments

Dipl.Ing. Nicoleta Popa[1]

1.   Introduction

The Research Fund for Coal and Steel is managed by the European Commission, Directorate-General for Research and Innovation, Directorate G (Industrial Technologies), Unit G.5. The types of actions supported are: Research Projects, Pilot and Demonstration Projects, Accompanying Measures.

Pilot and Demonstration projects are aiming to bridge the gap between Research and Innovation. Innovation can be considered as the technological implementation of new products or processes within the relevant industrial sector, or of significant improvements to products or processes, based on previous research results. The innovation is technologically implemented if it is introduced on the market (product innovation) or used within a production process (process innovation).

Demonstration projects aim at constructing and/or operating an industrial-scale installation or a significant part of an industrial-scale installation. Such projects aim to bring together all the technical and economic data in order to proceed with the industrial and/or commercial exploitation of the technology at minimum risk.

ECOBRIDGE project is a demonstration project funded by the Research Fund for Coal and Steel.

2.   Motivation

In the design and construction of bridges, questions of sustainability, maintenance and durability become more and more important for European road administrations in addition to safety and serviceability issues. From this perspective, integral abutment bridges turn out to become highly attractive to designers, constructors and road administrations. The main reason for this is that they tend to be less expensive to build, easier to maintain and more economical to own over their life time. This is principally due to the non-existence of bearings and joints that are main sources of maintenance costs during life time [1] [14].

---

1   Arcelor Luxembourg, Coordinator of the EcoBridge – RFCS Project

Furthermore the slender, robust, durable and proven WIB bridges (rolled beams in concrete) of the European rail systems can again be made competitive by using PreCoBeam (Prefabricated Composite Beams). By introducing these beams for example into framed load bearing systems such as integral abutment bridges, it is possible to manufacture hybrid integral structures with concrete shafts and composite crossbeams that can easily absorb the high horizontal forces from impacts or earthquakes. The fabrication of the beams in the factory and the reduction of the number of tasks on the building site lead to simple, effective quality assurance and improve health and safety protection during the building program [12].

However, in most European countries only little experience with integral bridges as well as PreCoBeam bridges, thus road administrations often are reluctant in using these innovative bridge types.

Therefore landmark projects are needed, aiming on presenting the knowledge gained within the scope of the INTAB (RFSR-CT-2005-00041) as well as the PRECOBEAM (RFSR-CT-2006-00030) projects, in a comprehensible way for the public as well as for authorities, designers and constructors.

Hence, the ECOBRIDGE project has as objectives the construction of three composite bridges with integral abutments and/ or PreCoBeam. The targeted countries are: Germany, Romania and Poland.

3. Background of the project and state of the art

Bridges are of vital importance to the European infrastructure and composite bridges already became a popular solution in many countries and a cost-effective and aesthetic alternative to concrete bridges [1][2][3][4]. Their competitiveness depends on several circumstances such as site conditions, local costs of material and staff and the contractor's experience. A major advantage is the savings in construction time, which reduces the traffic disturbance, consequently saves money for the contractor but even more for the road users; a fact that for a long time has been neglected. Recently this factor is increasingly drawn into focus as latest studies show the necessity of taking not only the simple production costs but also the construction time and the damage to national economy into account when deciding for a specific bridge type. [5][6][7]

Thus nowadays the following demands are made on bridge structures, which are all met by the construction of composite bridges [3][7].

- low production and maintenance costs;
- short construction time, to save costs for traffic control measures;
- construction of the bridge without essential interference of the traffic under the bridge;
- minimised traffic disturbance for maintenance.

All these needs have been proven to be met by integral abutment bridges [1][3], as this bridge type not only lowers production and maintenance costs but reduces economic and socio-economic costs as well.

- The superstructure can be designed quite slender, which decreases the construction height and the earthworks respectively. This leads to a decrease of material, fabrication, transport and construction costs.

- Frame bridges allow for the elimination of the middle support. This simplifies the construction of the bridge without essential interference of the traffic under the bridge, as the road has not to be closed.
- Due to the absence of bearings and joints, the maintenance costs can be decreased significantly.

Therefore bridges with integral abutments hold the potential to outclass traditional bridges with transition joints which has been proven clearly within the scope of the INTAB project [1].

Since 1998, bridges have been created in a composite pre-fabrication (VFT® = Verbund-Fertigteil-Träger = prefabricated composite beam) method of construction [11]. The system established with over 300 erected structures principally in Germany as well in Poland and Austria. It is a cost-effective construction for composite bridges of small and medium spans with site-prepared traffic deck.

The PRECOBEAM system – VFT girder with rolled girders in concrete – represents a further development of this method of construction. The new system provides a rolled beam section that is cut in the web centre in such a way to result in 2 T-sections, whereas the cutting form provides the shear connector. This special cut of steel web allows a perfect connection to the upper concrete part. The cutting guide selected for the manufacture of the concrete dowels enables the manufacture of tall sections without waste [1]. With the separation technology used it is possible to achieve a high quality for the separating faces with minimum local notch effects.

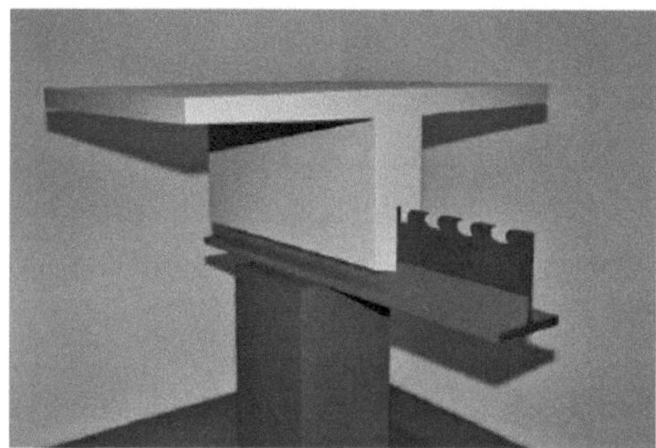

Figure 1:   PRECOBEAM girder

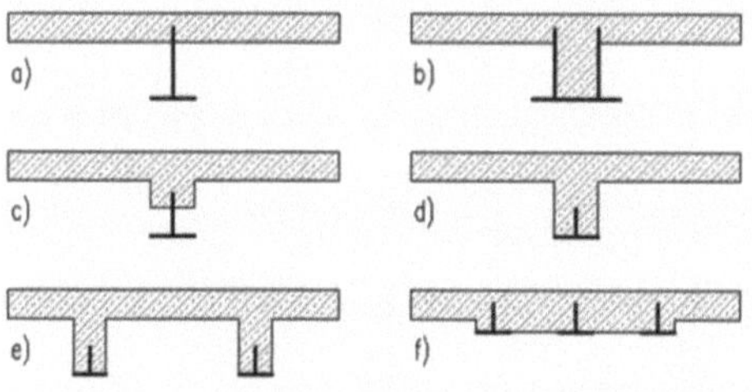

Figure 2:   Configuration variants of PRECO girders for bridge construction

The PRECO principle combines the advantages of the VFT girder with the robustness of the traditional "filler beam plate". The steel components consist of profiles with no upper chord as shown in the schematic representations in 2. The in-site concrete deck that is later completed is coupled by means of connecting reinforcement with the concrete chord of the pre-fabricated girder.

4.   Resources and partnership

The consortium members (involving three universities, three design offices, three steel contractors and one steel manufacturer) have long track record and extensive experience in this field, and complementary skills as necessary for successful completion of the project.

Universities

The **Institute for Steel Structures at Rheinisch-Westfälische Technische Hochschule (RWTH)** works in various fields with particular emphasis on the promotion of steel in the structural area. A particular research item is the interface between structural engineering and mechanical engineering for steel- and composite construction. In this field particular stability, composite action, use of high strength materials and fatigue aspects are investigated to develop economic and durable design rules. Other foci lie on dynamics of structures as well as log term and short term monitoring of engineering structures. RWTH has been the coordinator of INTAB project assuring the success of the project.

Wrocław University of Technology (PWr) with 32,000 students is one of the leading educational centres in Poland. The Faculty of Civil Engineering, with academic staff consisting of 169 people including 12 full professors and 17 associate professors, comprises three institutes and one of them is the Institute of Building Engineering. The Institute possesses an experimental research laboratory equipped with modern measurement devices. The institutes and their faculty divisions are deeply involved in different research programmes corresponding to their field of interests, among other things the field of composite structures is the subject. In 2004 research program concerning composite beams prestressed with external tendons have been conducted in the institute and within the confines of this program full scale experimental tests have been carried out.

The **University Politehnica Timisoara (UPT)**, participating to the project by the Department of Steel Structures and Structural Mechanics, chaired by Prof. D. Dubina, former president of ECCS, and member in TC7, TC8 and TC10 of ECCS. The research team from this university achieved significant reputation in the field of seismic resistant structures, participating to important European projects related to seismic resistant structures, like Copernicus "RECOS" and FP6 "PROHITECH", and presently is involved in RFCS "STEELRETRO" project. It has to be noted that the series of worldwide recognised STESSA conferences, devoted to Steel Structures in Seismic Areas, started in Timisoara in 1994.

Steel producer

ArcelorMittal Belval and Differdange S.A. – R&D department (APRL), located in Esch-sur-Alzette (Luxembourg), is the Research and Development Centre of the ArcelorMittal Long Carbon Steel sector, ArcelorMittal being one of the biggest steel producers worldwide.

APRL has an extensive experience in the steel research, especially in the scope of the F6 committee "Steel Structures" of ECSC and present TGS8 committee of RFCS, where it has been and is still for the time being involved in research dealing with fire safety engineering, design of steel and composite structures using normal and high strength materials, with EC1, EC3 and EC4 improvements, and many other fields. As a result of their contacts with their clients through the Trade, Marketing and Technical advice departments, they are aware of the needs of the construction world: architects, engineering offices, building contractors, students.

Engineering consultancies

Around 200 engineers, architects, designers and interdisciplinary specialists are working for **Schmitt Stumpf Frühauf and Partners, Engineering Consultancy (SSF)** in Germany. As consulting engineers for almost 30 years Mr. Schmitt, Mr. Stumpf and Mr. Frühauf have been designing and developing building, structures, tunnels and bridges with exacting requirements, performing structural investigations and feasibility studies. SSF has developed the Prefabricated Composite Girder method (VFT®) and built already more than 110 bridges with this construction method. Special areas here include the design of integral structures without joints and bearings in monolithic construction and the optimization of composite steel bridges. SSF has been the coordinator of PRECOBEAM project and was involved in INTAB project as an individual consultant.

    SSF Romania (SSF-RO) specializes in providing civil, structural and geotechnical engineering services for transport infrastructures, industrial facilities, buildings and properties. SSF has worked on many structures for highway and railway bridges in reinforced and pre-stressed concrete, steel and composite. SSF-RO has designed, inspected and/or refurbished bridges of virtually every class and balancing technically challenging designs with complex site considerations.

    Europrojekt Gdańsk Sp. zoo (Europrojekt) is a design-consulting company. Since the foundation in 1999 we have worked on numerous projects in Poland and Germany. The company is engaged in the design of roads, bridges, tunnels, public and utility facilities as well as commercial and residential buildings. Characteristic feature of the company is intensive development related to the improvement of the quality of our services and creating new job opportunities.

Building company / Beneficiary

TWT Sanierungsgesellschaft (TWT) is a construction company specialized in refurbishing and reconstructing existing structures. Founded 1993 the company engages about 20 site manager and technicians who are manufacturing steel constructions in the own workshop and implementing them on site. Furthermore concrete works are carried out also on site. On this basis of its industrial TWT is ideally suited for building composite constructions with a high standard of quality.

    Energopol-Szczecin SA (Energopol) is one of the biggest companies in the construction industry in north-western Poland. The company has extensive experience in the execution of construction works, including the construction of marine and inland hydraulic engineering (construction of ports, wharves, coastal fortifications, etc.), road-bridge construction, civil engineering, including sewage treatment plants, sewage collectors of large diameters, piling works for objects on land and water (bridges, overpasses, etc.), works relating to security of excavation in the form of palisades, steel sheet piling, strengthening of subsoil and concrete foundations.

5. Realized projects in Germany, Poland and Romania

Simmerbach Bridge, Germany

Simmerbach Bridge is part of the German railway network located in the Southwest of Germany. The requirement was to replace two single span bridges both with a span of 12,75m that were in operation for more than 100 years. The old bridges were designed as steel-constructions with a conventional ballast substructure. One of the two bridges crosses the Simmerbach River while the other one crosses a soil trail only. The bridges that had to be replaced are situated in a row so that the construction and replacement affects one among the two railway tracks only. On the opposite direction there are two bridges as well which were replaced about 30 years ago due to their bad condition at that time. They could be kept in operation and did not have to be either repaired or replaced.

Figure 3:   Simmerbach bridge – General view

The possibility to replace a bridge deck within less than 72 hours is a very important topic for railway bridges and certainly an important factor within the competition between different construction techniques.

PE4 bridge in Poland

The owner of the Polish bridge is General Directorate of National Roads and Motorways (GDDKiA) – the governmental institution administrating net of main Polish roads and motorways. For purposes of this project PreCoBeam technology was implemented to achieve better economic results – the idea was to use constant height of prefabricated girder but finally variable height of section at final stage.

The structural solution is a 4-span composite frame 17+22+22+17 m. It is realised by VFT technology and halved HEB1000/S355M sections, hence VFTWIB technology is used. For purposes of this project new technology (and modifications in VFTWIB technology) were implemented to achieve better economic results – the idea was to use constant height of prefabricated girder but finally variable height of section at final stage. It assumes additional calculation phase of girder (3 stages of concrete slab) but in fact the same number of steps at site – as additional slab (between prefabricated and final in-situ) is realised together with crossbeam. With this solution it was possible to achieve big lever arm at support to handle load of final in-situ slab and no cracks check at SLS was needed for reinforcement in this additional layer of concrete, so it could be highly tensioned. Variable constructional height resulted in elimination of influence of global action for design of steel dowels as level of shear connection is closed to level of neutral axis (at final stage).

Figure 4: PE4 bridge in Poland – PreCoBeam technology

Another 5 bridges have been built in Poland, proving that the PreCoBeam technology is a cost effective, environmental friendly and sustainable solution for bridges.

P11 bridge in Romania

The owner of the Romanian bridge is the Romanian National Company of Motorways and National Roads C.N.A.D.N.R. After a series of financial problems the Romanian team decided to enter the project also with a PreCoBeam solution. Through systemic analysis of the bearing capacity of composite dowel bars, they nowadays find a large application area in the field of bridge construction. A further development of the PreCoBeam (VFT®) construction method represents the development of the connection between concrete and steel eliminating the classical headed studs with concrete dowels; also the steel upper flange is detached. In this sense a new SSF system of precast composite girders was developed and introduced in Romania. This flexible system was conceived for middle spans and different situations including emergency cases; they have a very short erection time, small dead load and are very easy to handle.

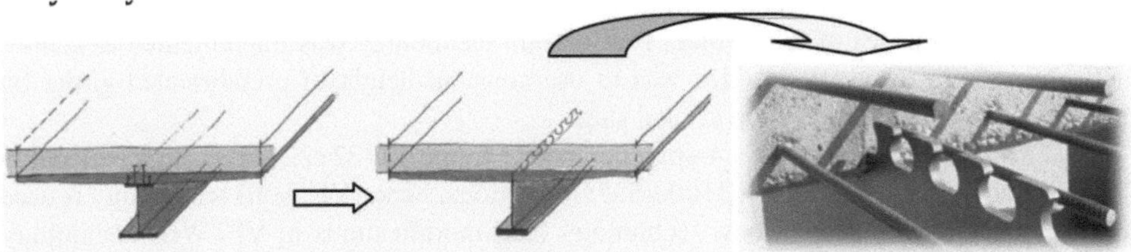

Figure 5:   Development of the composite cross sections

The Orăştie – Sibiu Motorway section has two structures (P01 and P11), constructed in the new PreCoBeam solution. This motorway is part of the corridor that crosses Romania from Nădlac to Constanţa and lies in the south of the Transilvanian basin, a natural depression formed in the north of the South Carpathians. This solution was adopted for two skew ≈ 70° overpasses, with one span of 36.5 ($\perp$) m. The girders are 39 m long and assure a clearance in the orthogonal direction of 35 m. Integral bridges are embedded on their abutments. A frame static system was adopted in this case as well.

SSF designed very economic composite integral overpasses using the clothoidal CL connection [15]. The novelty of the system represents the welded steel girders made of S355 J2+N from the prefabricated composite beams. The middle section of the steel girders is made of a bottom flange and a web with steel dowel cuttings on top. A very slender and aesthetic aspect was obtained for these overpasses and the solution allowed considerable material savings. Practically, for both bridges a very good steel consumption of 130 kg/m² was obtained.

Figure 6:   Integral overpasses over the A1 Motorway, section Orastie – Sibiu

6.   Conclusions

ECOBRIDGE proved to be a very successful project.

The efficient design and construction improves and consolidates the market position of the steel construction and steel producing industry. Additionally this advanced form of construction is contributing to savings in material and energy consumption for the structure during production and maintenance.

Steel technologies lead to benefits at a social and sustainability level, as they are fast to construct, highly pre-fabricated, flexible and adaptable in use, long life and recyclable.

The high degree of prefabrication helps to reduce the unintended manual works on-site leading to an upgrade of the working condition and quality of workmanship. Moreover the dangerous maintenance actions are largely reduced for bridges with integral abutments due to missing transition joints. Thus safer working conditions are achieved and remarkable saving in costs and down time are occurring for the lifetime of the bridge.

All time savings in construction and maintenance result in large benefits of the bridge owners but also for the community as less disturbance of the traffic will occur.

Additionally the exchange of technical experience from the partners from different parts of Europe contributes to a transfer of information from the different construction fields with different demands. Thus to perform this research proposal on the European level supports the idea of a higher degree of co-operation and harmonization in the European steel and construction sector.

References

[1]   PAK, D. et al. (2009), "Economic and Durable Design of Composite Bridges with Integral Abutments – Final Report RFSR-CT-2005-00041", RFCS publications, European Commission, Brussels (2009)

[2]   PAK, D. et al. (2008), "Integral Abutment Bridges", 5th European Conference on Steel and Composite Structures, Graz, 03.09.-05.09.2008, Brussels (BE): ECCS European Convention for Constructional Steelwork (2008), pp 189-194

[3]   COLLIN, P. (1996), "Some trends in Swedish Bridge Construction", International Conference on welded Structures, 2-3 September 1996, Budapest, Hungary, pp 163-172.

[4]   COLLIN, P. and JOHANSSON B. (1999), "Wettbewerbsfähige Brücken in Verbundbauweise", Stahlbau 68, Heft 11, pp 908-918.

[5]   KUHLMANN, U. et al. (2007) „Ganzheitliche Wirtschaftlichkeitsbetrachtungen bei Verbundbrücken unter Berücksichtigung des Bauverfahrens und der Nutzungsdauer" Stahlbau 76 (2007), Heft 2, pp 105-116

[6]   BRAUN, A., SEIDL, G., WEIZENEGGER, M. (2006) „Rahmentragwerke im Brückenbau", Beton- und Stahlbetonbau 101 (2006), Heft 3, pp 187-197

[7]   SCHMACKPFEFFER, H.. (1999), „Typenentwürfe für Brücken in Stahlverbundbauweise im mittleren Stützenbereich", Stahlbau 68 (1999), Heft 4, pp 264-276

[8]   NILSSON, M. (2008), "Evaluation of In-situ Measurements of Composite Bridge with Integral Abutments", Licentiate Thesis 2008:2, Luleå University of Technology (not published yet)

[9]   COLLIN, P., VELJKOVIC, M., PETURSSON, H. (2006), "International Workshop on the Bridges with Integral Abutments", Technical Report 2006:14, Lulea University of Technology, 2006

[10]   GLYNN, A. C., (2008), "Bridge Manual", New York State Department of Transportation, Office of Structures, 4th edition, 2008

[11]   SCHMITT, V., et alt.: VFT-Bauweise, Entwicklung von Verbundfertigteilträgern im Brückenbau, Beton- und Stahlbetonbau 96, 2001, Heft 4

[12]   SEIDL, G. et al. (2006), "Prefabricated Enduring Composite Beams based on Innovative Shear Transmission – Proposal RFSR-CT-2006-00030 "

[13]   NILSSON, M. et. al. (2008), "Towards a better understanding of behaviour of bridges with integral abutments", Construction Conference VI, Colorado 2008

[14]   FELDMANN, M. et. al. (2009), "Composite bridges with integral abutments", 8th Japanese German Bridge Symposium, 03.-05.08.2009, Munich, Germany

[15]   SEIDL, G., PETZEK, E., BANCILA, R., Composite Dowels in Bridges – Efficient solution, International Conference ISCS13, Advanced Materials Research, ISBN-13: 978-3-03785-848-6, 2013.

The right choice of steel – according to the Eurocode

Oliver Hechler [1], Georges Axmann & Boris Donnay [2]

Keywords: steel, production, steel grade, material properties, ductility, toughness, weldability.

Abstract: In general, the choice of the steel grade is ruled in Eurocode EN 1993-1-1. Several requirements are specified: choice according to the material properties, ductility requirements, toughness properties and through-thickness properties. With reference to these requirements on the mechanical characteristics, modern hot-rolled structural sections are produced by precise control of the temperature during the rolling process. Fine grain steels, produced using thermomechanical rolling (delivery condition M according to EN 10025-4), feature improved toughness values which give a lower carbon equivalent and a fine microstructure when compared with conventional or normalised steels. This paper gives guidance on and background to the right choice of the steel grade according to the Eurocode. Furthermore, the influence of the production process on this choice is highlighted and the advantages of thermomechanical steels for each criterion are discussed.

1.   Introduction

Eurocode 3 [1] applies to the design of buildings and civil engineering works in steel. It complies with the requirements and principles for the safety and serviceability of structures, the basis of their design and verification that are given in EN 1990 – Basis of structural design. Requirements are provided for resistance, serviceability, durability and fire resistance of steel structures. These are based on the principle of limit state design, which mainly assumes that the resistance of cross-sections and members specified for the ultimate limit states are based on tests in which the tolerances are met according to EN 1090-2 [2], and the material exhibited sufficient ductility to apply simplified design models. Therefore, the material properties, for steel the steel grade, have to be specified in detail to comply with the safety level of Eurocode 3 ("Fig. 1").

These simplified design models and the safety concept of the Eurocode are based on tests at ambient temperature, for which ductile failure occurs as the steel is on the upper shelf region with sufficient toughness. In Fig. 2 (left), the conclusions from testing for the partial safety factors and the characteristic strength are shown. If brittle fracture takes place, the assumptions for the design models and the safety concepts are no longer met ("Fig. 2", right). Consequently, failure against brittle fracture must be accounted for with an appropriate choice of steel with sufficient toughness.

---

1   ArcelorMittal Long Carbon Europe, Technical Advisory, Luxembourg
2   ArcelorMittal Belval and Differdange, R&D, Long Products, Luxembourg

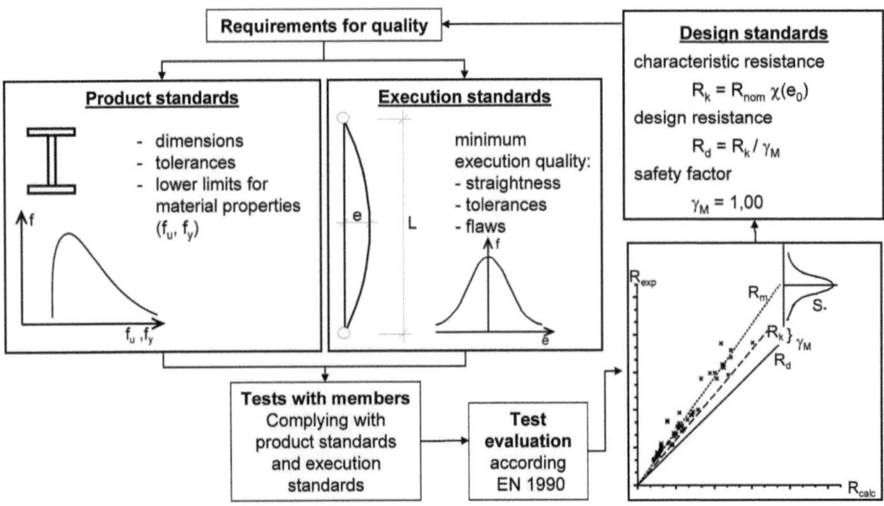

Figure 1:   Reliability of strength verification [3]

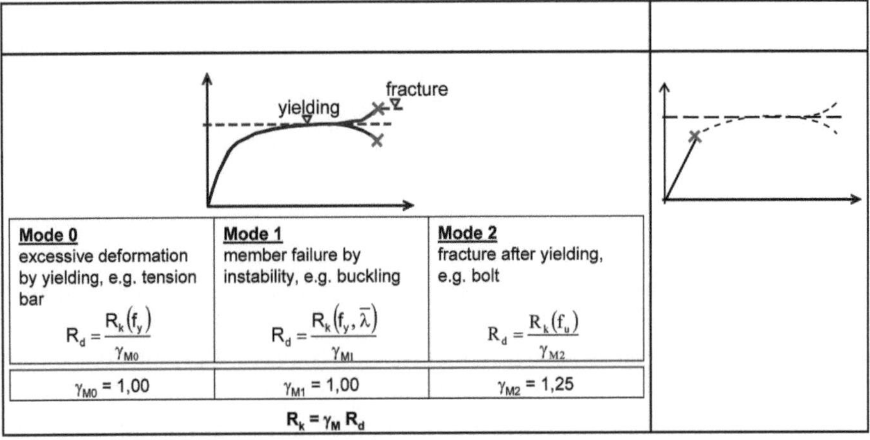

Figure 2:   Brittle fracture and ductile failure [3]

2.   The choice of the steel grade

In general, the choice of the steel grade is ruled in Eurocode EN 1993-1-1 (2005). Several requirements are specified:

- Choice according to the mechanical material properties
  Nominal values of material properties are defined as characteristic values in design calculations.
- Ductility requirements
  For steels, a minimum ductility is required.
- Toughness properties

Simplified aids are given to choose the appropriate material with sufficient fracture toughness to avoid brittle fracture.
- Through-thickness properties.
  Guidance on the choice of through-thickness properties is given in EN 1993-1-10 (2005).

With reference to these requirements, the designation of the steel grade is defined in the product standard for hot-rolled products and structural steels in EN 10025 (2004) ("Fig. 3"). The classification of steel grades is accordingly based on the minimum specified yield strength at ambient temperature.

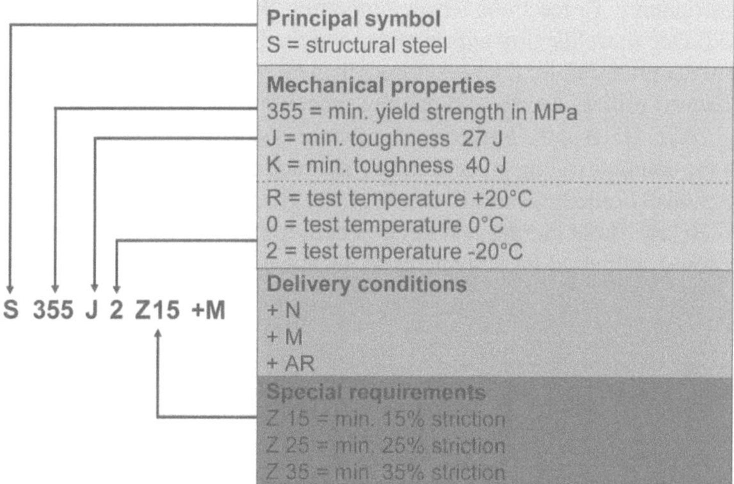

Figure 3:   Designation of the steel grade according to EN 10025 (2004)

The product standard also differentiates the delivery condition. These are defined by the rolling process of the steel. Conventional hot rolling takes place in the recrystallised austenitic phase (γ) and is followed by a subsequent air cooling, see Fig. 4. Without any special rolling control or heat treatment, this material delivery condition is specified to be "as rolled" (AR) in the EN 10025. With an additional normalising thermal treatment (N), the steel microstructure can be refined leading to improved properties specifically described in EN 10025-3, if alloying elements have been added to the steel. The reduction of the grain size leads to an increase of the specific surface of the grain boundaries within the material. Since these grain boundaries represent an obstacle to deformation the yield strength increases. The fine grained structure of normalised steels may further be improved with restriction of the alloying elements via a thermomechanical treatment. This may be quenching (Q) with water or oil, an accelerated cooling to about 500°C after rolling and lastly slower cooling to the room temperature, and a successive tempering (T) to regain ductility. A fine grained microstructure may also be obtained if the hot rolling process is carried out with a control of the temperature during the final deformation (CR). Another possibility to refine the microstructure is to apply a thermomechanical rolling process according to EN 10025-4. Hereby, rolling is also performed with a controlled rolling process in the recrystallised austenitic and further rollings in the non-crystallised austenitic phase, in cases even into the austenitic-ferritic phases ($\alpha + \gamma$) (M/N). For thermomechanical steels (TM), rolling is carried out at lower temperatures than normalising rolling; the rolling temperature in the finishing stand is typically close to the transformation temperature of the austenite ($\gamma_{\text{non-recr.}}$ ® $\alpha + \gamma$). The grain size of austenite is about 20 µm or larger before the last rolling passes. After rolling, the austenite grains are usually elongated because of the sluggish recrystallization of the microstructure due to the low rolling temperature.

Although controlled rolling leads to an attractive combination of strength and ductility, it also includes substantial disadvantages. The reduction of the rolling temperature brings an increase of the rolling loads and many mills are not designed to resist the additional stresses. Because a waiting time is usually incorporated in the rolling schedule, controlled rolling can increase rolling time and adversely reduce productivity. Moreover, with higher material thickness, the rolling temperature increases and the air cooling rate after rolling decreases, which induces rougher microstructures. To reach the tensile properties, the content of alloying elements has to be adapted. Due to weldability requirements and the limit in equivalent carbon, beams in grade S460 are not produced for thickness larger than 50 mm. To overcome the limitations of thermomechanical rolling, accelerated cooling process of beams after rolling has been developed (TM + QST). Hereby, the fine grained structure is achieved by a minimum of alloying elements with the complex rolling process and a strict temperature control. As the ferrite grain size of conventionally rolled steels is 10 to 30 µm, the grain size of TM + QST steels is usually between 5 to 10 µm. These fine grained steels benefit from a low carbon equivalent value and are to be predominantly used for large material thicknesses in high strength steel to address weldability.

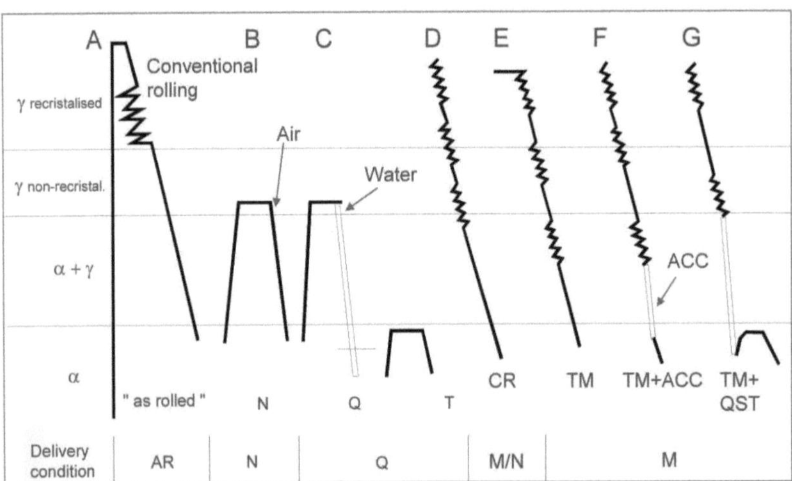

Figure 4:   Relation of the delivery condition to the rolling process

In addition to the group of thermomechanical steels delivered according to EN 10025-4, ArcelorMittal has developed steel grades to fully valorise the potential of quenching and self-tempering (QST) process. These fine grained TM-steels are branded HISTAR® steels [Z-30.2-5] and are characterised by more stringent requirements in terms of mechanical properties and chemical composition.

2.1   Mechanical properties

The nominal values of the yield strength f_y and the ultimate strength f_u for structural steel should be obtained by adopting the values $f_y = R_{eh}$ and $f_u = R_m$ direct from the product standard, see Tab. 1, or by a Tab. drafted from this standard in EN 1993-1-1. It has to be noticed, that the required yield strength decreases with increasing material thickness. This takes into account the effect, that with the increase in material thickness, the addition of alloying elements need to be higher to achieve constant yield strength over the thickness. However, with the increase in addition of alloying elements, the carbon equivalent value raises and welding becomes problematic. Welding is substantial to the application of structural steels. Thus, the normative rules have considered this fact by lowering the required yield strength for thicker plates to account for weldability.

Table 1:   Mechanical properties at ambient temperature for thermomechnical rolled steels [4]

Designation		Minimum yield strength R_{eH} [a] MPa[b]						Tensile strength R_m [a] MPa[b]					Minimum percentage elongation after fracture[c] % $L_0 = 5{,}65\ \sqrt{S_0}$
		Nominal thickness mm							Nominal thickness mm				
According EN 10027-1 and CR 10260	According EN 10027-2	≤ 16	> 16 ≤ 40	> 40 ≤ 63	> 63 ≤ 80	> 80 ≤ 100	> 100 ≤ 120 [d]	≤ 40	> 40 ≤ 63	> 63 ≤ 80	> 80 ≤ 100	> 100 ≤ 120 [d]	
S275M S275ML	1.8818 1.8819	275	265	255	245	245	240	370 to 530	360 to 520	350 to 510	350 to 510	350 to 510	24
S355M S355ML	1.8823 1.8834	355	345	335	325	325	320	470 to 630	450 to 610	440 to 600	440 to 600	430 to 590	22
S420M S420ML	1.8825 1.8836	420	400	390	380	370	365	520 to 680	500 to 660	480 to 640	470 to 630	460 to 620	19
S460M S460ML	1.8827 1.8838	460	440	430	410	400	385	540 to 720	530 to 710	510 to 690	500 to 680	490 to 660	17

[a] For plate, strip and wide flats with widths ≥ 600 mm the direction transverse (t) to the rolling direction applies. For all other products the values apply for the direction parallel (l) to the rolling direction.

[b] 1 MPa = 1 N/mm²

[c] For product thickness < 3 mm for which test pieces with a gauge length of L_0 = 80 mm shall be tested, the values shall be agreed at the time of the enquiry and order.

[d] For long products a thickness ≤ 150 mm applies.

The producers verify the conformity of their products with the standard by tensile tests, in which for each section, the location of the test specimen is also defined, see e.g. Fig. 5 for beams. The result of the tensile test is the stress-strain curve from which the relevant parameters, yield strength f_y and tensile strength f_u, are determined. These parameters are exemplarily indicated in Fig. 6, a typical stress-strain curve for a HISTAR®460 (or S460 steel grade according to EN 10025-4 for thermomechanical rolled weldable fine grain structural steels).

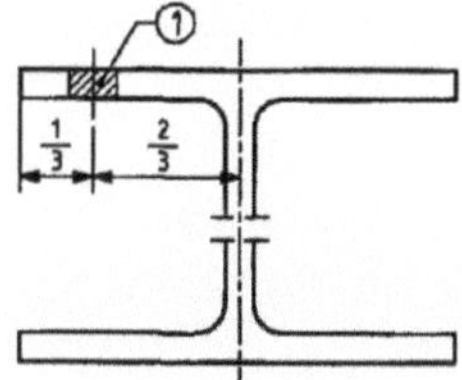

Figure 5:   Location of test specimen for tensile test

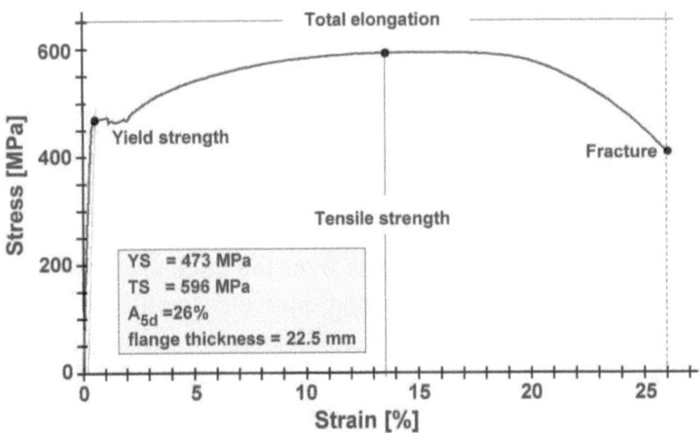

Figure 6:   Stress-Strain diagram from a HISTAR®460 steel

For thermomechanical rolled fine grained steels of the new generation using the QST process (e.g. HISTAR® steels), it is remarkable that a decrease of the yield strength in respect to the material thickness can be avoided without an increase of the alloying elements and the carbon equivalent value. A comparison of the material thickness to yield strength in relation to steels according to EN 10025 (2004) and modern HISTAR® steels according to Z-30.2-5 (2008) is given in Fig. 7. As a result, the right choice of thermomechanical steel gives the designer an economical advantage in design, as presented in the example of application of this paper.

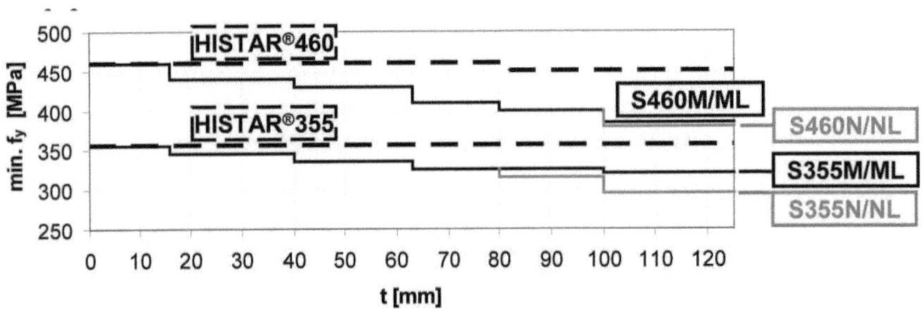

Figure 7:   Comparison of the material thickness t to yield strength f_y in relation to steels according EN 10025 (2004) and modern HISTAR® steels according to Z-30.2-5 (2008)

2.2   Ductility

Ductility is required to avoid brittle failure of structural elements. For steels, a minimum ductility is required that should be expressed in terms of limits for:

- the elongation at failure on a gauge length of $5.65\sqrt{A_0}$ (where A_0 is the original cross-sectional area); Eurocode recommends an elongation at failure not less than 15%;
- the ratio f_u / f_y of the specified minimum ultimate tensile strength f_u to the specified minimum yield strength f_y; Eurocode recommends a minimum value of $f_u / f_y \geq 1.10$.

Both criteria are of particular interest for high strength structural steels as the grade S460 due to the fact that the higher the yield strength, the less elongation will be present at failure ("Fig. 8"). The minimum required elongation for structural steels is given in Tab. 1. Therefore, the product standard offers more ductility than required in EN 1993-1-1. However, Fig. 5 also illustrates that the minimum required elongation is in general met with a high safety margin by modern steels of higher strength. The ratio f_u / f_y is in general more critical than the minimum elongation. Therefore, various tensile test have been compiled and the ratio f_y / f_u has been plotted over the yield strength ("Fig. 9").

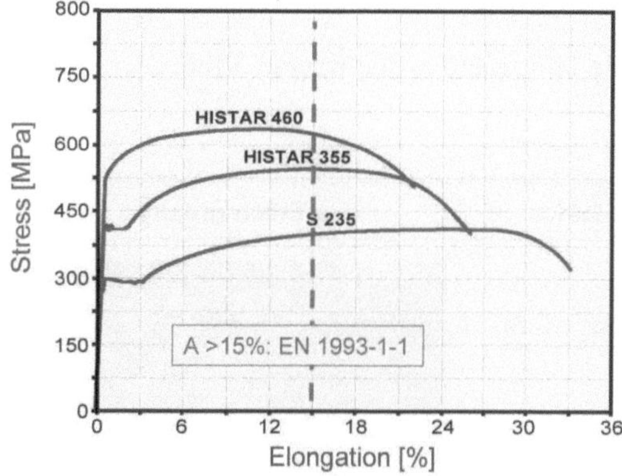

Figure 8:   Comparison of stress-strain curves for S235 to S335 and S460 steel of the modern generation

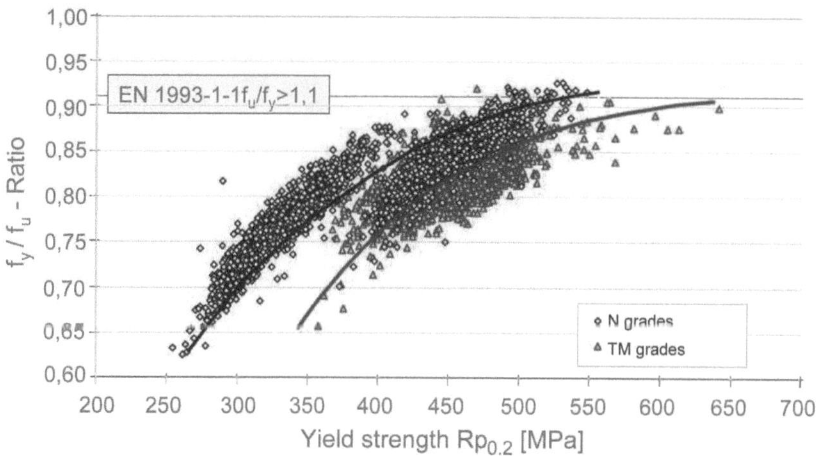

Figure 9:   Ratio of yield strength to tensile strength for structural steels of ArcelorMittal

The conclusion from the diagram is that structural steels up to 460 MPa fulfil the ductility criteria. Structural steels with yield strengths higher than 460 MPa seem, on the first look, not to be able to fulfil the ductility criteria. Thermomechanical steels are well adopted to fulfil these criteria with thanks given to their specific strengthening mechanism (refined microstructure and reduced microalloying content).

In the Hong Kong Code of Practice for the Structural Use of Steel (2005) [5], a ratio f_u / f_y = 1.2 is required and therefore 9% more conservative compared to the Eurocode and therefore does not allow for high strength, high toughness steels. Further, the elongation at failure to be not less than 15% is required which is in line with the Eurocode requirement. The ration 1.2 used in the Hong Kong Steel Code is reasonable for conventional steels.

2.3   *Toughness*

2.3.1   Introduction

There are two ways of material failure: ductile failure and brittle fracture (Tab. 2).

Table 2:   Failure mechanisms of materials

Failure mode	Deformation of crystal lattice	Fractography
Ductile failure • shear • slipping • toughness • dull		
Brittle fracture • cleavage • decohesion • brittleness • shiny		

Toughness is the resistance of a material to brittle fracture when stressed. Toughness is defined as the amount of energy per volume that a material can absorb before rupturing. The material toughness depends on:

- Temperature
  Materials lose their crack resistance capacity with decreasing temperature ("Fig. 10"). This relation can be displayed in an impact energy A_v – temperature T curve with an upper shelf region (3: ductile failure), lower shelf region (1: brittle fracture) and a transition region (2: crack shows shares of cleavage and shear area).
- Influence of loading speed
  The higher the loading speed, the lower the toughness ("Fig. 11").
- Grain size
  The orientation of the crystal lattice varies in the adjacent grains ("Fig. 12"). Whenever the crack tip reaches the grain boundary, the crack would subsequently change his

growth direction and thus energy is dissipated. Consequently, fine grained steels are more resistant to brittle failure.

- Cold forming
  With an increase in cold forming, the yield strength increases with decreasing ductility ("Fig. 13").
- Material thickness
  In the two dimensional stress state, steel plastic deformation starts at the yield point. In the three-dimensional stress state, the crystal lattice of the steel is compacted from all sides and therefore the steel yield strength is increased significantly. Thus, thinner plates with a higher share of material in the two-dimensional stress state do have more ductility than thicker plates ("Fig. 14").

The material toughness is in general experimentally investigated by the Charpy impact test with the resulting impact energy – temperature curve.

Further relevant factors which have also an influence on the resistance of members to brittle fracture are:

- Notch detail
  Crack initiation highly depends on the notch detail and the resulting stress, crack position and crack shape expressed by the notch intensity factors ("Fig. 15").
- Load utilisation level of member
  The higher the tension in the member, the higher the failure probability ("Fig. 16").

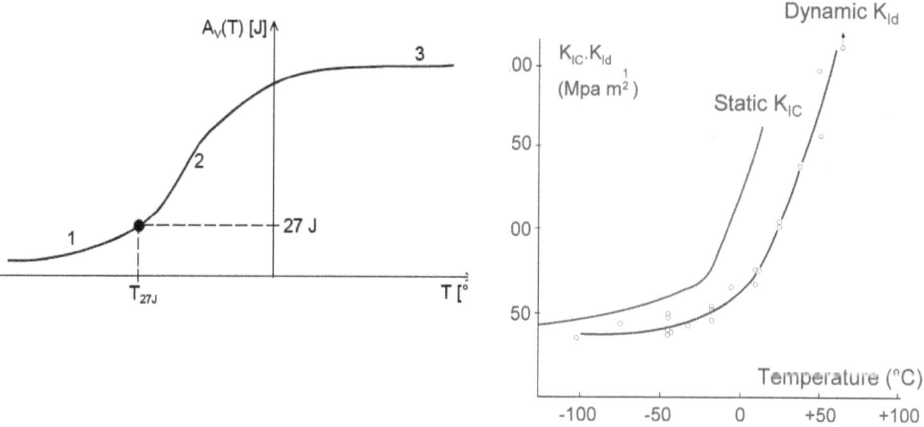

Figure 10: Impact energy A_v – temperature T curve

Figure 11: Stress intensity factor – temperature curve for quasi-static and dynamic loading

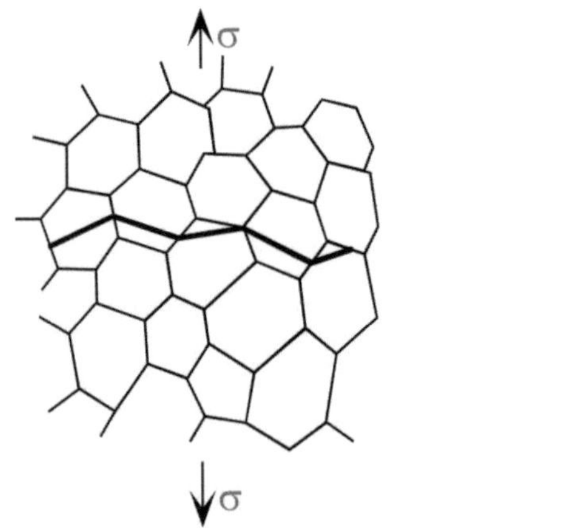

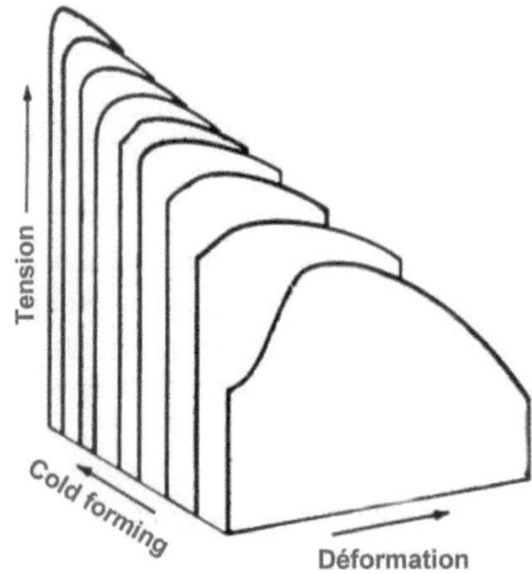

Figure 12: Model of crack propagation in the crystal lattice

Figure 13: Stress-strain curve in dependency of the degree in cold forming

Figure 14: Fracture surfaces of Charpy impact tests for plates with different material thicknesses

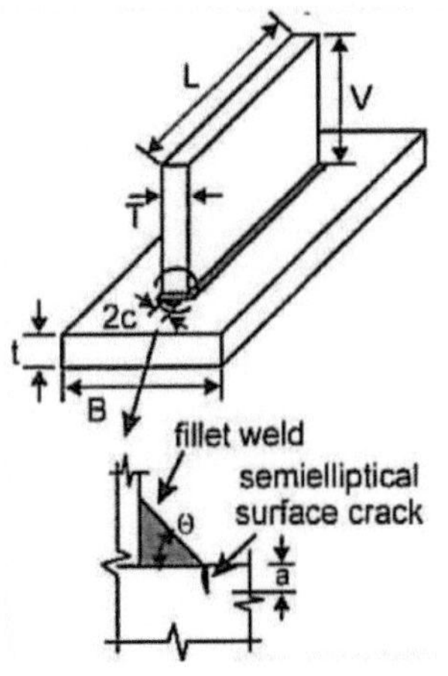

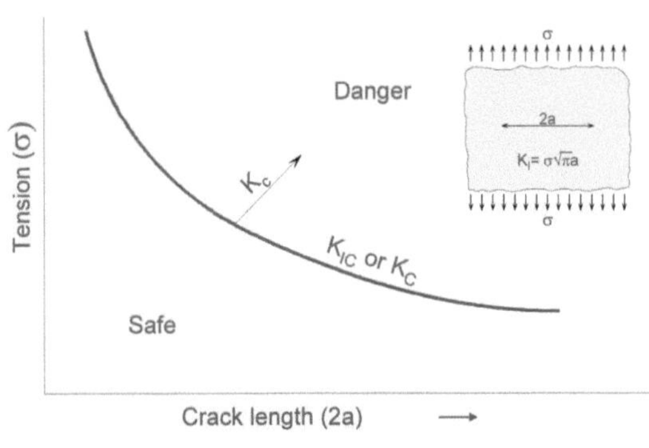

Figure 15: Specification of a notch for the determination of the notch intensity factor

Figure 16: Relation of the failure loading to the crack length

2.3.2   The right choice of steel for toughness assessment

As introduced, it is essential for the safe use of structures designed according to Eurocode 3 that the material has a sufficient toughness to avoid brittle fracture of tension elements exposed at the lowest service temperature expected to occur within the intended design life of the structure. Rules for the selection of steels are therefore offered in Part 1-10 of EN 1993, which allow a simple toughness assessment to avoid brittle fracture. The basis for the assessment is a fracture mechanics approach with the design check for which the design values of the action effect $E_d = K^*_{appl,d}$ (stress intensity factor) are compared to the design values of the toughness resistance $R_d = K_{mat,d}$ in the transition region of the impact energy – temperature curve Kühn (2005) ("Fig. 17") [6] thus as:

$$K^*_{appl,d} \leq K_{mat,d} \tag{1}$$

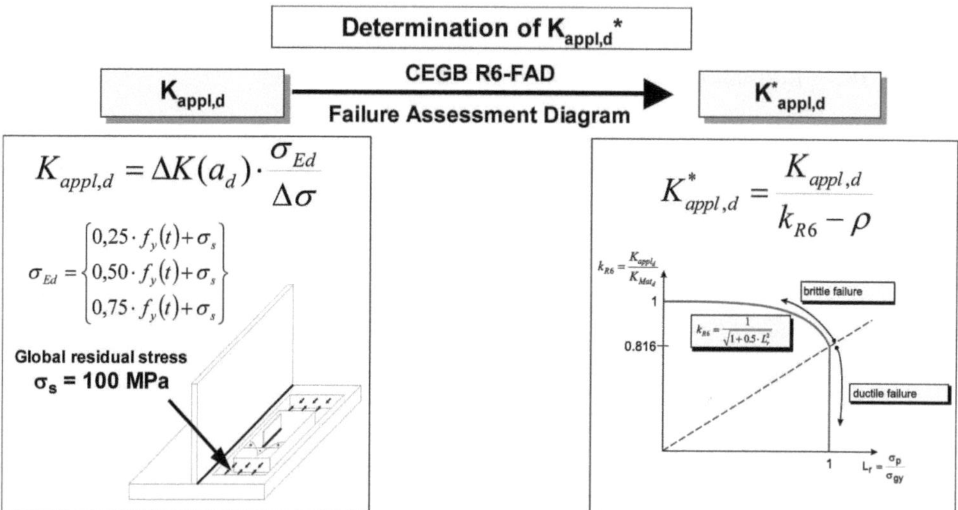

Figure 17:   Action side of the toughness assessment [7]

On the action side, $K_{appl,d}$ is determined for a certain flaw size, modeled by a surface crack. However, the use of the stress intensity factor $K_{appl,d}$ is limited for an elastic fracture mechanics approach and therefore needs to be corrected for plastic strains via the CEGB R6-FAD Failure Assessment Diagram. Consequently, the toughness requirement for the steel is given by the $K^*_{appl,d}$.

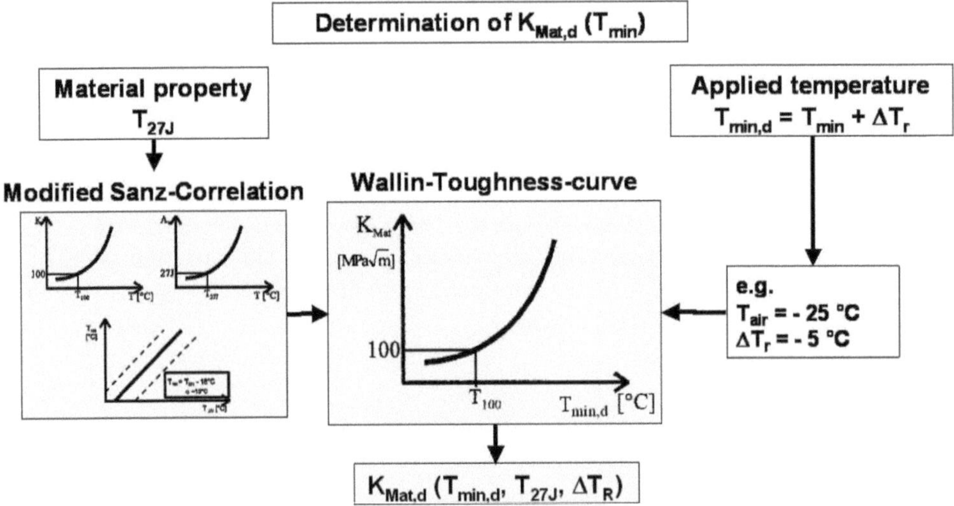

Figure 18:   Resistance side of the toughness assessment [7]

On the resistance side, the temperature for 27J derived from the Charpy impact test is transferred with the modified Sanz-Correlation ($T_{K100} = T_{27J} - 18°C$) in the T_{K100} temperature as input value for the Wallin-Toughness-curve. From this curve, the threshold value for brittle fracture $K_{mat,d}$ is finally derived for the assessment ("Fig. 18").

$$K^*_{appl,d} = \frac{K_{appl,d}}{k_{R6} - \tilde{n}} \leq 20 + \left[70 \left\{ \exp \frac{T_{Ed} - T_{27J} + 18°C + \Delta T_R}{52} \right\} + 10 \right] \left(\frac{25}{b_{eff}} \right)^{0.25} \tag{2}$$

A transformation in the temperature format needs to be carried out:

$$T_{Ed} - 52 \cdot \ln \frac{\left(K^*_{appl,d} - 20 \right) \left(\dfrac{b_{eff}}{25} \right)^{0.25} - 10}{70} + \Delta T_R \geq T_{27J} - 18°C \tag{3}$$

Resulting in the design check format as per EN 1993-1-10 for toughness requirement:

$$T_{Ed} \geq T_{Rd} \tag{4}$$

Hereby, all relevant parameters from the action and from the structure are included in T_{Ed}, whereas T_{Rd} contains the material properties from the tests only. The results of this assessment have been summarized and are presented in the EN 1993-1-10 by a Tab. for permissible material thickness for the choice of the steel grade ("Tab. 3"). This Tab. represents therefore a simple design aid for the practising engineer.

Input parameters are hereby:

- reference temperature T_{ed} in °C which can also be used to consider toughness reducing effects due to cold forming, impact loading by a fictive transformation of the reference temperature ($T^*_{Ed} = T_{Ed} - \Delta T_{Ed}$);
- the stress state σ_{Ed} assumes to occur simultaneously with the reference temperature; the steel grade in accordance to the delivery condition.

With the double checking according to Fig. 2: on one side, the appropriate choice of the steel grade for the reference temperature in relation to the stress state to be in the transient region of the impact energy – temperature curve of the toughness; and on the other side, the resistance of the members according to EC3-1-1 in the upper region of the impact energy – temperature curve the safety requirements according to the Eurocodes, are both satisfied. As alternative methods to the simplified toughness assessment are presented above, the fracture mechanics method (in this method the design value of the toughness requirement should not exceed the design value of the toughness property) as well as the numerical evaluation (this may be carried out using one or more large scale test specimens) may be used in the Eurocode.

Table 3: Simplified method for the determination of permissable material thicknesses for standardized details, Tab. 2.1 of [8]

Steel grade	Sub-grade	Charpy energy CVN at T [°C]	J_{min}	$\sigma_{Ed} = 0,75\,f_y(t)$ 10	0	-10	-20	-30	-40	-50	$\sigma_{Ed} = 0,50\,f_y(t)$ 10	0	-10	-20	-30	-40	-50	$\sigma_{Ed} = 0,25\,f_y(t)$ 10	0	-10	-20	-30	-40	-50
S235	JR	20	27	60	50	40	35	30	25	20	90	75	65	55	45	40	35	135	115	100	85	75	65	60
	J0	0	27	90	75	60	50	40	35	30	125	105	90	75	65	55	45	175	155	135	115	100	85	75
	J2	-20	27	125	105	90	75	60	50	40	170	145	125	105	90	75	65	200	200	175	155	135	115	100
S275	JR	20	27	55	45	35	30	25	20	15	80	70	55	50	40	35	30	125	110	95	80	70	60	55
	J0	0	27	75	65	55	45	35	30	25	115	95	80	70	55	50	40	165	145	125	110	95	80	70
	J2	-20	27	110	95	75	65	55	45	35	155	130	115	95	80	70	55	200	190	165	145	125	110	95
	M,N	-20	40	135	110	95	75	65	55	45	180	155	130	115	95	80	70	200	200	190	165	145	125	110
	ML,NL	-50	27	185	160	135	110	95	75	65	200	200	180	155	130	115	95	230	200	200	200	190	165	145
S355	JR	20	27	40	35	25	20	15	15	10	65	55	45	40	30	25	25	110	95	80	70	60	55	45
	J0	0	27	60	50	40	35	25	20	15	95	80	65	55	45	40	30	150	130	110	95	80	70	60
	J2	-20	27	90	75	60	50	40	35	25	135	110	95	80	65	55	45	200	175	150	130	110	95	80
	K2,M,N	-20	40	110	90	75	60	50	40	35	155	135	110	95	80	65	55	200	200	175	150	130	110	95
	ML,NL	-50	27	155	130	110	90	75	60	50	200	180	155	135	110	95	80	210	200	200	200	175	150	130
S420	M,N	-20	40	95	80	65	55	45	35	30	140	120	100	85	70	60	50	200	185	160	140	120	100	85
	ML,NL	-50	27	135	115	95	80	65	55	45	190	165	140	120	100	85	70	200	200	200	185	160	140	120
S460	Q	-20	30	70	60	50	40	30	25	20	110	95	75	65	55	45	35	175	155	130	115	95	80	70
	M,N	-20	40	90	70	60	50	40	30	25	130	110	95	75	65	55	45	200	175	155	130	115	95	80
	QL	-40	30	105	90	70	60	50	40	30	155	130	110	95	75	65	55	200	200	175	155	130	115	95
	ML,NL	-50	27	125	105	90	70	60	50	40	180	155	130	110	95	75	65	200	200	200	175	155	130	115
	QL1	-60	30	150	125	105	90	70	60	50	200	180	155	130	110	95	75	215	200	200	200	175	155	130
S690	Q	0	40	40	30	25	20	15	10	10	65	55	45	35	30	20	20	120	100	85	75	60	50	45
	Q	-20	30	50	40	30	25	20	15	10	80	65	55	45	35	30	20	140	120	100	85	75	60	50
	QL	-20	40	60	50	40	30	25	20	15	95	80	65	55	45	35	30	165	140	120	100	85	75	60
	QL	-40	30	75	60	50	40	30	25	20	115	95	80	65	55	45	35	190	165	140	120	100	85	75
	QL1	-40	40	90	75	60	50	40	30	25	135	115	95	80	65	55	45	200	190	165	140	120	100	85
	QL1	-60	30	110	90	75	60	50	40	30	160	135	115	95	80	65	55	200	200	190	165	140	120	100

As mentioned previously, fine grained steels as the HISTAR® thermomechanical grades exhibit a high toughness and therefore are perfectly suited for use in heavy sections.

The Code outlines the necessity to prevent brittle fracture. Similar to the Eurocode, the assessment procedure of the Code is also based on fracture mechanics approach – however, the assessment itself is more refined. In the simplified approach of the Eurocode, the toughness requirement is determined for longitudinal attachments, as this is known as the most severe detail. In the Code, the maximum allowable material thickness is assessed from the maximum basic thickness (for a specified minimum service temperature, 27J Charpy impact value and the strength grade of steel) multiplied by a K-factor (for type of joint detail, stress level and strain conditions). Therefore, the toughness assessment according to the Code may benefit from the consideration of the structural details investigated – however, advantages of higher toughness addressed by the Eurocode (subgrades, "Tab. 3") stay disregarded.

2.3.3   Through thickness properties

Lamellar tearing is a type of weld-cracking that occurs beneath a weld ("Fig. 19"). It may form when certain plate materials presenting low ductility in the thickness (or through) direction are welded to a perpendicular element. The failure by tearing is generally located within the base metal outside the heat-affected zone and parallel to the weld fusion boundary. The problem is caused by welds that the base metal is subjected to high shrinkage stresses in the thickness direction. The main parameter governing the deformation behavior in its through-thickness direction on the material side is the sulphur, contained as a residual element in the steel. However, it is known that only the deformation behavior and not the strength in through thickness direction can be improved by the steel manufacturing process.

Therefore, if necessary, lamellar tearing is avoided by the choice of the base material with adequate ductility in the thickness direction. This choice defines the quality class for through-thickness properties according to EN 10164 (2004) [8], the Z-grade as special requirement in the steel designation (Z 15, Z 25, Z 35) ("Fig. 3"), and should be selected depending on the consequences of lamellar tearing.

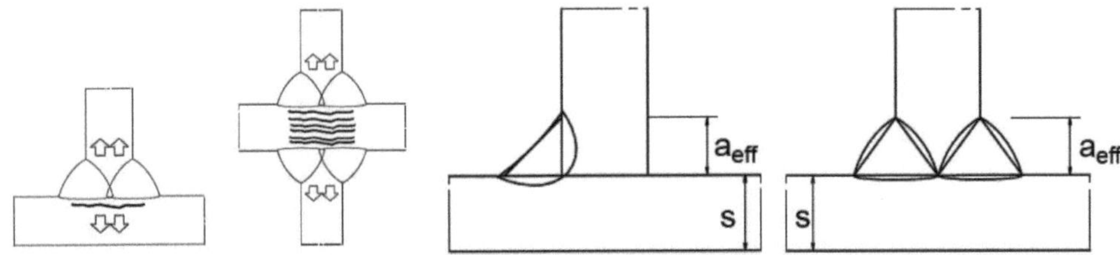

Figure 19:   Lamellar tearing [9] **Figure 20:**   Effective weld depth a_eff for shrinkage, Fig. 3.2 of [9]

The following aspects should be considered in the selection of steel assemblies or connections to safeguard against lamellar tearing ("Tab. 4"):

- The criticality of the location in terms of applied tensile stress and the degree of redundancy.
- The strain in the through-thickness direction in the element to which the connection is made. This strain arises from the shrinkage of the weld metal as it cools. It is greatly increased where free movement is restrained by other portions of the structure.
- The nature of the joint detail, in particular the welded cruciform, tee and corner joints.
- Chemical properties of transversely stressed material. High sulphur levels, in particular, even if significantly below normal steel product standard limits, can increase the risk of lamellar tearing.

The susceptibility of the material should be determined by measuring the through-thickness ductility quality to EN 10164, which is expressed in terms of quality classes identified by Z-values. Lamellar tearing may be neglected if the following condition is satisfied according to EN 1993-1-10 (2005):

$$Z_{Ed} \geq Z_{Rd} \tag{5}$$

where Z_{Rd} is the available design Z-value for the material according to [8] and Z_{Ed} is the required design Z-value determined using:

$$Z_{Ed} = Z_a + Z_b + Z_c + Z_d + Z_e \tag{6}$$

in which Z_a, Z_b, Z_c, Z_d and Z_e are as given in Tab. 4 and a_{eff} according to Fig. 20.

The Code also addresses the properties of steel in the direction perpendicular to the product surface. In the Code, it is stipulated that for design stresses in through-thickness exceeding 90%, steel with guaranteed through-thickness properties shall be specified.

Consequently, lamellar tearing is in general not an issue in design if materials with an appropriate through-thickness quality produced with modern production techniques are chosen and if a rational design of the welds has been carried out. Further, the improved through-thickness properties are generally appreciated to mitigate the risk of lamellar tearing during fabrication only (which may typically occur after welding of heavily restrained welds on thick products) and to a lesser extent to mitigate the risk of fracture in service (e.g. columns of moment frames in seismic conditions).

Table 4:   Criteria affecting the target value of zed, Tab. 3.2 of [9]

a)	Weld depth relevant for straining from metal shrinkage	Effective weld depth a_{eff} (see Figure 3.2) = throat thickn. a of fillet welds		Z_i
		$a_{eff} \leq 7$mm	$a = 5$ mm	$Z_a = 0$
		$7 < a_{eff} \leq 10$mm	$a = 7$ mm	$Z_a = 3$
		$10 < a_{eff} \leq 20$mm	$a = 14$ mm	$Z_a = 6$
		$20 < a_{eff} \leq 30$mm	$a = 21$ mm	$Z_a = 9$
		$30 < a_{eff} \leq 40$mm	$a = 28$ mm	$Z_a = 12$
		$40 < a_{eff} \leq 50$mm	$a = 35$ mm	$Z_a = 15$
		$50 < a_{eff}$	$a > 35$ mm	$Z_a = 15$
b)	Shape and position of welds in T- and cruciform- and corner-connections	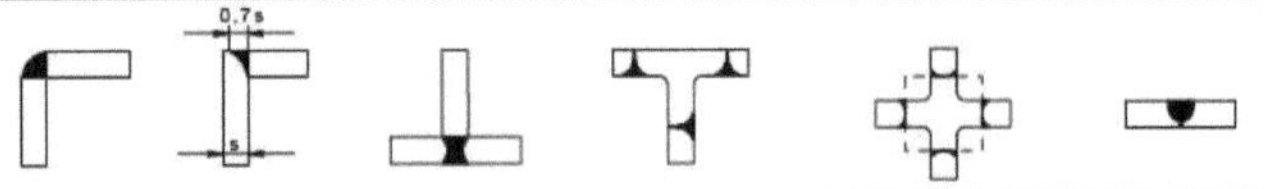		$Z_b = -25$
		corner joints	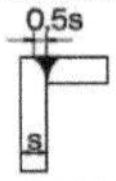	$Z_b = -10$
		single run fillet welds $Z_a = 0$ or fillet welds with $Z_a > 1$ with buttering with low strength weld material	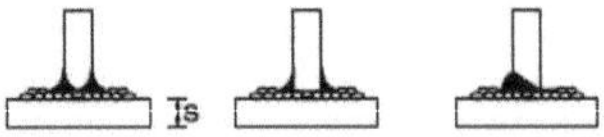	$Z_b = -5$
		multi run fillet welds		$Z_b = 0$
		partial and full penetration welds	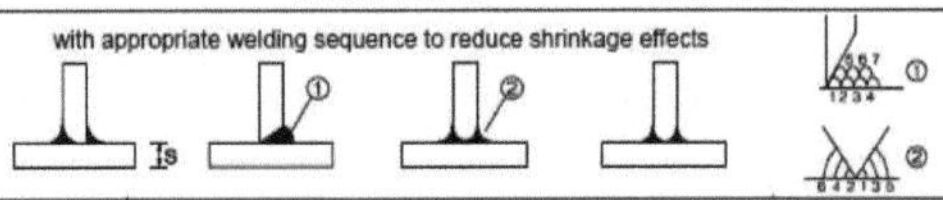 with appropriate welding sequence to reduce shrinkage effects	$Z_b = 3$
		partial and full penetration welds	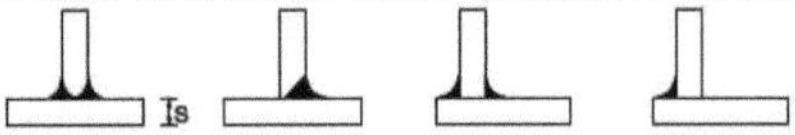	$Z_b = 5$
		corner joints		$Z_b = 8$
c)	Effect of material thickness s on restraint to shrinkage	$s \leq 10$mm		$Z_c = 2^*$
		$10 < s \leq 20$mm		$Z_c = 4^*$
		$20 < s \leq 30$mm		$Z_c = 6^*$
		$30 < s \leq 40$mm		$Z_c = 8^*$
		$40 < s \leq 50$mm		$Z_c = 10^*$
		$50 < s \leq 60$mm		$Z_c = 12^*$
		$60 < s \leq 70$mm		$Z_c = 15^*$
		$70 < s$		$Z_c = 15^*$
d)	Remote restraint of shrinkage after welding by other portions of the structure	Low restraint:	Free shrinkage possible (e.g. T-joints)	$Z_d = 0$
		Medium restraint:	Free shrinkage restricted (e.g. diaphragms in box girders)	$Z_d = 3$
		High restraint:	Free shrinkage not possible (e.g. stringers in orthotropic deck plates)	$Z_d = 5$
e)	Influence of preheating	Without preheating		$Z_e = 0$
		Preheating $\geq 100°C$		$Z_e = -8$

* May be reduced by 50% for material stressed, in the through-thickness direction, by compression due to predominantly static loads.

3.   Weldability of modern steel grades

The weldability of steels highly depends on the hardenability of the steel, which is an indication of the prosperity to form martensite during cooling after heat treatment. The hardening of steel depends on its chemical composition, with greater quantities of carbon and other alloying elements resulting in a higher hardenability and thus a lower weldability. In order to be able to compare alloys made up of many distinct materials, a measure known as the equivalent carbon content (CEV) is used to evaluate the relative weldability of different alloys. As the equivalent carbon content rises, the weldability of the steel decreases. There is a trade-off between material strength and weldability: low alloy steels are characterised by a reduced resistance and higher alloying contents by a poor weldability. However, with the thermomechanical rolling process, high strength steel can be produced without substantial increase in the carbon equivalent and therefore, keeping an excellent weldability even for thick products. A comparison of the resulting carbon equivalent value for a normalised, thermomechanical rolled and HISTAR®steel over the material thickness is shown in Fig. 21.

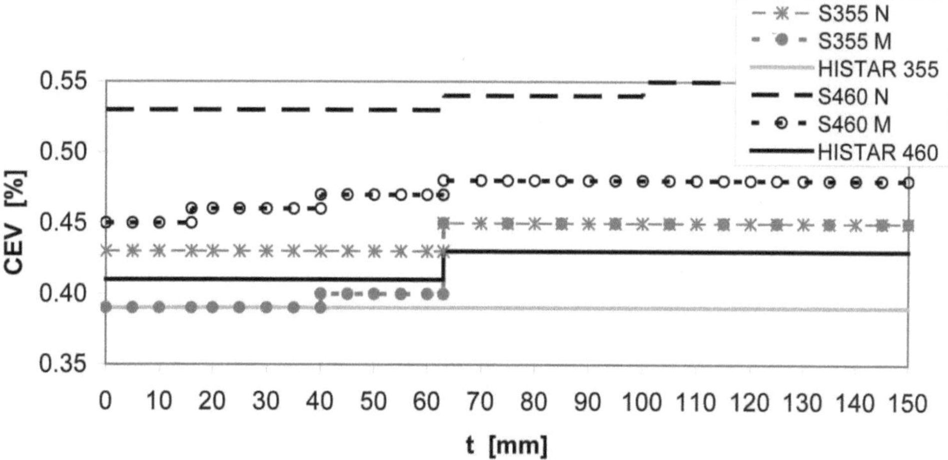

Figure 21:   Comparison of the maximum CEV for nominal product thickness acc. to [4] with CEV of the HISTAR® steels

Hereby, the resulting carbon equivalent values have been calculated as follows:

$$CEV = C + Mn / 6 + (Cr + Mo + V) / 5 + (Ni + Cu) / 15 \qquad (5)$$

Consequently, thermomechanical rolled steels with a reduced carbon equivalent value do simplify the welding. The forming of martensite, however, can be avoided using preheating which increases the t8/5-time during welding but decreases the productivity of the welding workshop. Due to the low carbon equivalent value of HISTAR® steels, welding of structural sections with flange thicknesses up to 125mm thickness is even possible without preheating in ambient temperature > 0°C ("Fig. 22").

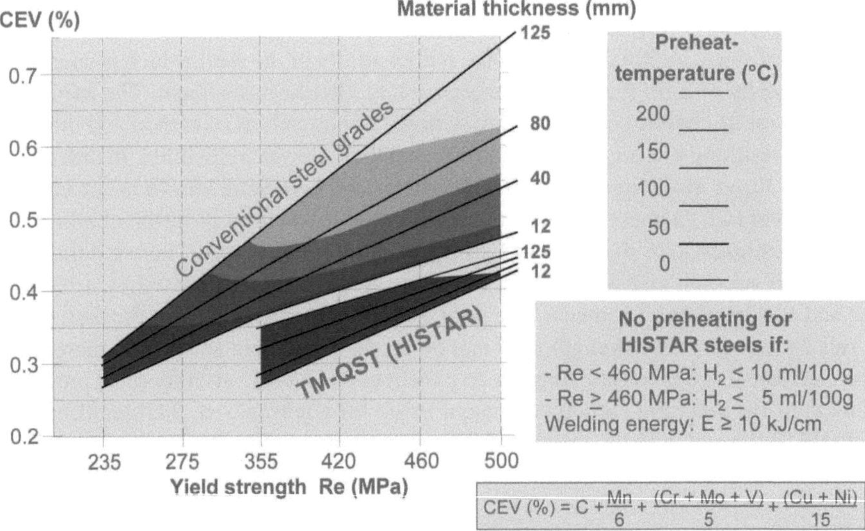

$$CEV\ (\%) = C + \frac{Mn}{6} + \frac{(Cr + Mo + V)}{5} + \frac{(Cu + Ni)}{15}$$

Figure 22:   Weldability and preheat temperatures [10]

As a result, a remarkable gain in economy is achieved. For example, the welding of a column with an HD 400 x 1086 section with a flange thickness of tf = 125 mm and a web thickness of tw = 78 mm, 140 passes are required for a full penetration weld, talking about 8 hours of welding (Fig. 23). For example, for normalized S355 steel, preheating is required to 110°C which consumes an additional 4 hours. With the use of thermomechanical steel in HISTAR® quality by ArcelorMittal, consequently 1/3 of the welding time can be saved. Further welding becomes much more convenient for the welder.

Figure 23:   Welding of a column HD 400 x 1086 kg/m in HISTAR® 460 (tf = 125 mm, tw = 78 mm)

Weldability is also addressed in the Code by limiting the carbon content to 0.24%, the sulphur and phosphor content to 0.03% and the carbon equivalent value to 0.48%. For steels according to the EN 10025-4 (2004) and HISTAR®, these requirements are satisfied and they can therefore be used.

4. Further fabrication of modern fine grain steels

European recommendations about shaping of structural steels are given in document TR 10347 (2006) [11]. In essence, the following is mentioned therein:

- Cold forming
  Structural steel grades can be cold-formed regardless whether their delivery condition AR, N, or M. Cold forming decreases ductility of steel and increases its yield strength (valid for all grades), see Fig. 13. Straining requirements for certain bending radius/ thickness and the plate thickness are stipulated in EN 1993-1-8.
- Hot-forming
  M grades are not dedicated to hot-forming. Only N grades can be hot formed. N grades have to be renormalized after hot forming. Hot forming (rerolling, hot bending) is not economical and very rarely applied to long products (i.e. beams).
- Heat straightening
  Heat straightening is used in fabrication of steel to remove or give a structural component a certain shape. It is carried out by a fast and short heating of the steel locally with oxy-acetylene burners. If needed, to increase the effectiveness of the process, additional restraint is applied, i.e. by means of hydraulic jacks, clamping or a dead load. Recommended maximum flame straightening temperatures to be respected are according to Tab. 5, Tab. 2 from [9]. The Tab. shows that up to grade S460, the same maximum value of flame-straightening temperature for N (normalized) steels as for M (thermomechanical) steels applies.

Table 5:   Recommended maximum values of the flame-straightening temperature, Tab. 2 from [11]

Delivery condition	Recommended maximum values of the flame-straightening temperature		
	Short superficial heating °C	Short full section heating °C	Full section heating with longer holding time °C
normalized	≤ 900	≤ 700	≤ 650
thermomechanical rolled up to S460	≤ 900	≤ 700	≤ 650
thermomechanical rolled S500 to S700	≤ 900	≤ 600	≤ 550
quenched and tempered	≤ tempering temperature applied to the original product – 20 K (generally below 550 °C)		

- Post weld heat treatment (PWHT)
  In addition to CEN TR 10347, it should be noted that if stress relieving (PWHT) is required, which is generally uncommon in fabrication of structural steel shapes, the same recommendations apply to structural steels, regardless of the delivery conditions AR, N and M. The generally recommended PWHT procedures are to apply temperatures between 530 to 580 °C and a holding time of 2 minutes per mm product thickness, but not less than 30 minutes and not more than 90 minutes

With regard to their technological properties, the thermo-mechanically rolled steels have therefore good cold-forming properties. Similar to conventional structural steels, they can be flame straightened provided specific maximum temperatures are not exceeded. In case stress relieving is considered for reducing residual stresses, the usual parameters concerning temperature range and heating time according to the rules of practice must be applied. Hot-forming, which is however uncommon for the fabrication of sections, must not be performed.

5.   Example of the appliance of modern steels in high rise buildings

The advantages of the application of HISTAR® high strength steels can be perfectly demonstrated on two similar buildings in the Spanish city of Barcelona ("Fig. 24").

Both 40-storey buildings consist of the same structural concept, similar dimensions and a comparable base grid. As a result, the values of the column loads resulting from dead load and life loads are comparable for the same type of column of every building.

While the structural steel construction of the first erected building is fabricated based on S 355 steel grade, the later erected second building is made of HISTAR®460. The result is significant: For the highest loaded columns in the lower part of the building, the column weight of the HISTAR®460 columns is reduced by 28% compared to those of S 355.

The higher yield strength of the HISTAR®460 steel is not the only reason for the same column capacity of the lighter section: Because of the controlled fabrication process resulting in lower imperfections, EN 1993-1-1 (2005) classifies hot rolled sections in S460 / HISTAR®460 in buckling curve a or a0 (Fig. 25), depending on the sections dimension. The second effect is through reduced dimensions and plate thicknesses due to higher allowable tresses, S 460 / HISTAR®460 are already in the more favourable range for the buckling curves definition.

Figure 24:   Hotel Olympia (S 355) and Mapfre Tower (HISTAR®460) in Barcelona, Spain

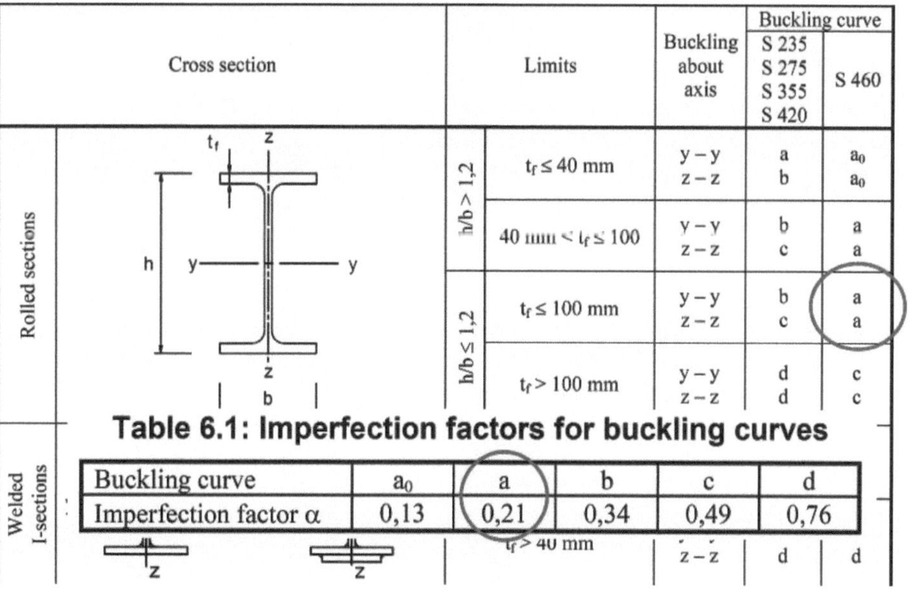

Cross section		Limits	Buckling about axis	Buckling curve S 235 S 275 S 355 S 420	S 460
Rolled sections	$h/b > 1{,}2$	$t_f \le 40$ mm	$y-y$ $z-z$	a b	a_0 a_0
		40 mm $< t_f \le 100$	$y-y$ $z-z$	b c	a a
	$h/b \le 1{,}2$	$t_f \le 100$ mm	$y-y$ $z-z$	b c	a a
		$t_f > 100$ mm	$y-y$ $z-z$	d d	c c
Welded I-sections		$t_f > 40$ mm	$z-z$	d	d

Table 6.1: Imperfection factors for buckling curves

Buckling curve	a_0	a	b	c	d
Imperfection factor α	0,13	0,21	0,34	0,49	0,76

Figure 25:   Choice of column buckling curves [1]

These effects give the most favourable results with the biggest economical advantages for high loaded columns with usual buckling length of 3-4m (one storey floor to floor height). For the above given example, globally 24% weight reduction for all columns are achieved ("Fig. 26").

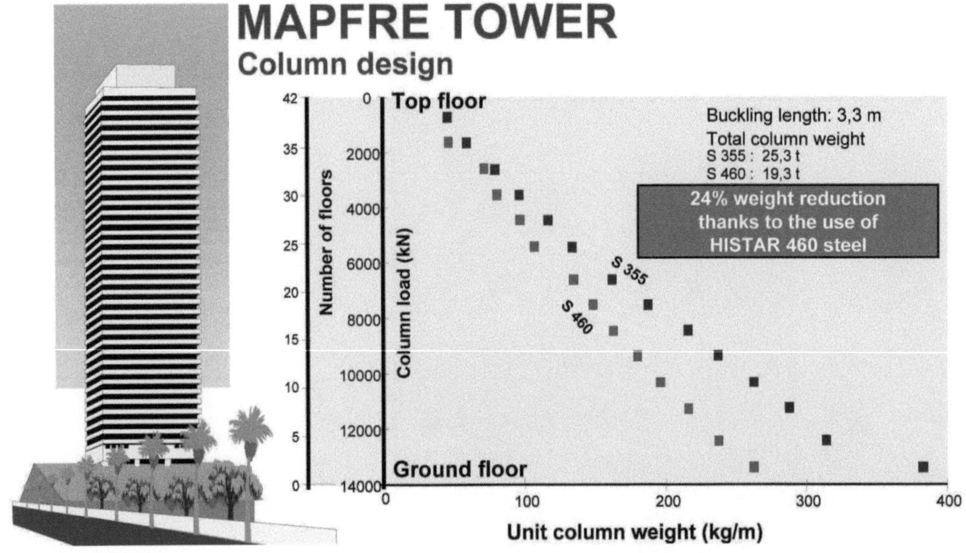

Figure 26:   Column weight comparison of S355 to HISTAR®460 for the Mapfre Tower

6.   Summary

The European practice for the right choice of the steel grade is ruled in Eurocode EN1993-1-1, which defines requirements on the mechanical material properties, ductility, toughness properties and through-thickness properties. In this paper, the requirements have been described and discussed. With reference to these requirements, modern hot-rolled structural sections, which are produced by precise control of the temperature during the rolling process, have been introduced. These steels, produced using thermo-mechanical rolling according to EN 10025-4 (2004), feature improved toughness values which give lower carbon equivalent values and a fine grained microstructure when compared with normalised steels. Further, high performance steel grades are available, e.g. the HISTAR® by ArcelorMittal. The high yield strength, good toughness at low temperatures and excellent weldability of HISTAR® steels make them a cost-effective choice, in particular for the design of multi-storey buildings, large span trusses and heavily loaded industrial construction. Compared with basic steels, these exceptional mechanical properties can reduce the construction weight by 25-50 %, depending on structural layout and can provide high strength and exceptional durability.

References

[1]   EN 1993-1-1. Eurocode 3: Design of steel structures – Part 1-1: General rules and rules for buildings, Brussels, Belgium, 2005

[2]   EN 1090-2. Execution of steel structures and aluminium structures – Part 2: Technical requirements for the execution of steel structures, Brussels, Belgium, 2004

[3]   Schäfer et al, "Modern Plastic Design for Steel Structures (PLASTOTOUGH)", Final Report, Research Programme of the Research Fund for Coal and Steel, RFSR-CT-2005-00039. 2005 – 2008, not yet published, 2009

[4]   EN 10025. Hot rolled products of structural steels, Brussels, Belgium, 2004

[5]   Code of practice for the structural use of steel, The Government of the Hong Kong Special Administration Region, Building Department, 2005

[6]   Sedlacek, G., Höhler, S., Kühn, B.,Langenberg, P., "Bruchmechanische Methoden im Sicherheitssystem des Stahlbaus", Stahlbau 72,Germany, 2003

[7]   Sedlacek, G.; Müller, C., " The use of high strength steels in metallic construction", 1st International Conference SUPER HIGH STRENGTH STEELS, Rome, Italy, 2005

[8]   EN 10164. Steel products with improved deformation properties perpendicular to the surface of the product – Technical delivery conditions, Brussels, Belgium, 2004

[9]   EN 1993-1-10. Eurocode 3: Design of steel structures – Part 1-10: Material toughness and through thickness properties, Brussels, Belgium, 2005

[10]   EN 1011-2. Welding – Recommendation for welding of metallic materials – Part 2: Arc welding of ferritic steels, Brussels, Belgium, 2001

[11]   prCEN/TR 10347. Guidance for forming of structural steels in processing, Brussels, Belgium, 2006

[12]   Deutsches Institut für Bautechnik, Allgemeine Bauaufsichtliche Zulassung Z-30.2-5. Langerzeugnisse aus warmgewalzten schweißgeeigneten Feinkornbaustählen im thermo-mechanisch gewalzten Zustand HISTAR 355/355L and HISTAR 460/460L, Berlin, Germany, 2008

[13]   Kühn, B., Beitrag zur Vereinheitlichung der europäischen Regelungen zur Vermeidung von Sprödbruch, Schriftenreihe Stahlbau – RWTH Aachen, Heft 54, Shaker Verlag, Aachen, Germany, 2005

[14]   G. Sedlacek, M. Feldmann, B. Kühn, D. Tschickardt, S. Höhler, C. Müller, W. Hensen, N. Stranghöner, W. Dahl, P. Langenberg, S. Münstermann, J. Brozetti, J. Raoul, R. Pope, F. Bijlaard, COMMENTARY AND WORKED EXAMPLES to EN 1993-1-10 "Material toughness and through thickness properties" and other toughness oriented rules in EN 1993. JRC Scientific and Technical Reports, Italy, 2008

Sustainable bridges – LCA for a composite and a concrete bridge

AL. Hettinger[1], JP. Birat[1], O. Hechler[2] & M. Braun[2]

Keywords: Life Cycle Assessment, bridges, composite bridges, steel production, emission.

Abstract: Life Cycle Assessment addresses the environmental impacts of a product's life cycle, from raw material extraction through production, use and end-of-life. In this paper, we apply this standardized method to two bridges: one designed as a composite bridge and one as a prestressed concrete bridge. The environmental profile of the bridges, defined by eleven indicators, is strongly connected to the bill of quantities. As a result, the composite bridge generates significantly less environmental impacts than its equivalent made of prestressed concrete, much heavier. The study also demonstrates that recycling is not necessarily beneficial depending on the material. On the one hand, the recycling of structural steel avoids the emission of 32 tCO2eq, thus decreasing the overall impact of the composite bridge, and, on the other hand, reinforced concrete requires a pre-treatment before recycling that is not counterbalanced by the benefits of recycling, thus downgrading the overall environmental profile of the prestressed bridge.

1.   Introduction

This paper presents the results of the Life Cycle Assessment (LCA) carried out on two bridges: one designed as a composite bridge and one as a prestressed concrete bridge. An LCA is a standardised methodology that addresses "the environmental aspects and potential environmental impacts throughout a product's life cycle from raw material acquisition through production, use, end-of-life treatment, recycling and final disposal" [1] Because of this holistic approach, LCA is a powerful tool to evaluate environmental aspects of the built environment. A change in product design will indeed affect different phases of the life cycle (recyclability, scarcity of resources…) and the life cycle approach will help understand the distribution of the burdens among the life cycle stages and avoid overcharging one stage by a change of design.

The LCA of the bridges has been conducted according to this standard and reviewed by external experts [2]. This paper first briefly describes the two bridges and then develops their LCA, following the four stages described in the standard ("Fig. 1").

---

1   ArcelorMittal Global Research & Development
2   ArcelorMittal Long Carbon Europe – Technical Advisory

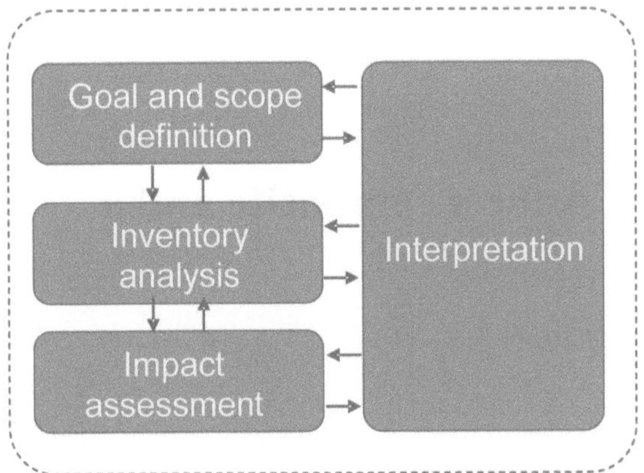

Figure 1:   Steps of a Life Cycle Assessment as defined by ISO standards [1].

2.   Description of the bridges

The two bridges exhibit a typical layout with two spans of 2 x 29.27m. They were designed to fulfil the same function. Fig. 2 shows the example of a typical two-span composite bridge.

Figure 2:   Composite bridge over A5 highway at Bremgarten (D) with 2 x 30m span, rolled beams in HISTAR460 and concrete crossbeams

Two alternatives have been compared: solution A, a prestressed concrete bridge, and solution B, a composite bridge made of a steel bearing structure and partially prefabricated concrete elements. The section of this latter bridge is shown in Fig. 3 and the bill of quantities for both solutions is shown in Fig. 4.

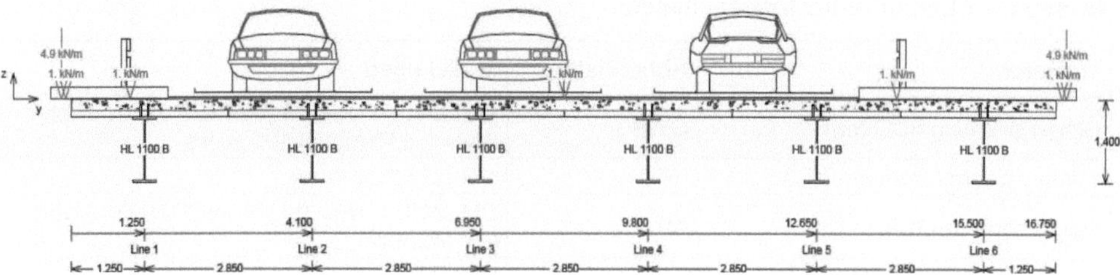

Figure 3:   Cross section of the composite bridge

A. Prestressed concrete		B. Composite bridge	
Concrete	2 926 t	Concrete	860 t
Reinforcing steel	145 t	Reinforcing steel	47 t
Prestressing steel	40 t	Sections	139 t
		End plates	4 t
		Shear studs	2 t
Total A	3 111 t	Total B	1 052 t

Figure 4:   Bill of quantities of the two solutions

3.   Description of the LCA

The evaluation aims at determining the environmental burdens generated by the bridge life cycle and comparing the results.

The environmental impacts of the bridges are defined in this paper by eleven parameters, which describe the environmental burdens generated by the bridges like global warming or toxicity.

These indicators, described in Tab. 1, are common LCA indicators. We have used the methods of the Institute of Environmental Sciences (CML) [3] to evaluate them, except for primary energy demand and water consumption which directly stem from mass and energy balances.

Table 1:   List of indicators evaluated

Indicator	Abbreviation	Model used
Global Warming (tCO2eq)	GWP	CML2001 – Nov. 09 [GUINEE 2001], Global Warming Potential 100 years
Primary Energy Demand (GJ)	PED	Net calorific values
Acidification (tSO2eq)	AP	CML2001 – Nov. 09 [GUINEE 2001], Acidification Potential
Eutrophication (tPO4eq)	EP	CML2001 – Nov. 09 [GUINEE 2001], Eutrophication Potential
Photochemical Ozone Creation (tC2H2eq)	POCP	CML2001 – Nov. 09 [GUINEE 2001], Photochem. Ozone Creation Potential
Ozone Depletion (tR11eq)	ODP	CML2001 – Nov. 09 [GUINEE 2001], Ozone Layer Depletion Potential
Water consumption (t)	WCP	None
Freshwater Toxicity (tDCBeq)	FTP	CML2001 – Nov. 09 [GUINEE 2001], Freshwater Aquatic Ecotoxicity Pot.
Human Toxicity (tDCBeq)	HTP	CML2001 – Nov. 09 [GUINEE 2001], Human Toxicity Potential
Marine Toxicity (tDCBeq)	MTP	CML2001 – Nov. 09 [GUINEE 2001], Marine Aquatic Ecotoxicity Pot.
Terrestrial Toxicity (tDCBeq)	TTP	CML2001 – Nov. 09 [GUINEE 2001], Terrestrial Ecotoxicity Potential

The system considered here is based on the life cycle of the bridges ("Fig. 5") that comprises several processes:

- The production phase of steel and concrete elements, from raw materials extraction to the production of finished products
- Transportation of steel and concrete elements to the construction site
- Erection of the bridge
- Use phase
- Dismantling
- Transportation to end-of-life treatment sites
- End-of-life (EOL) phase

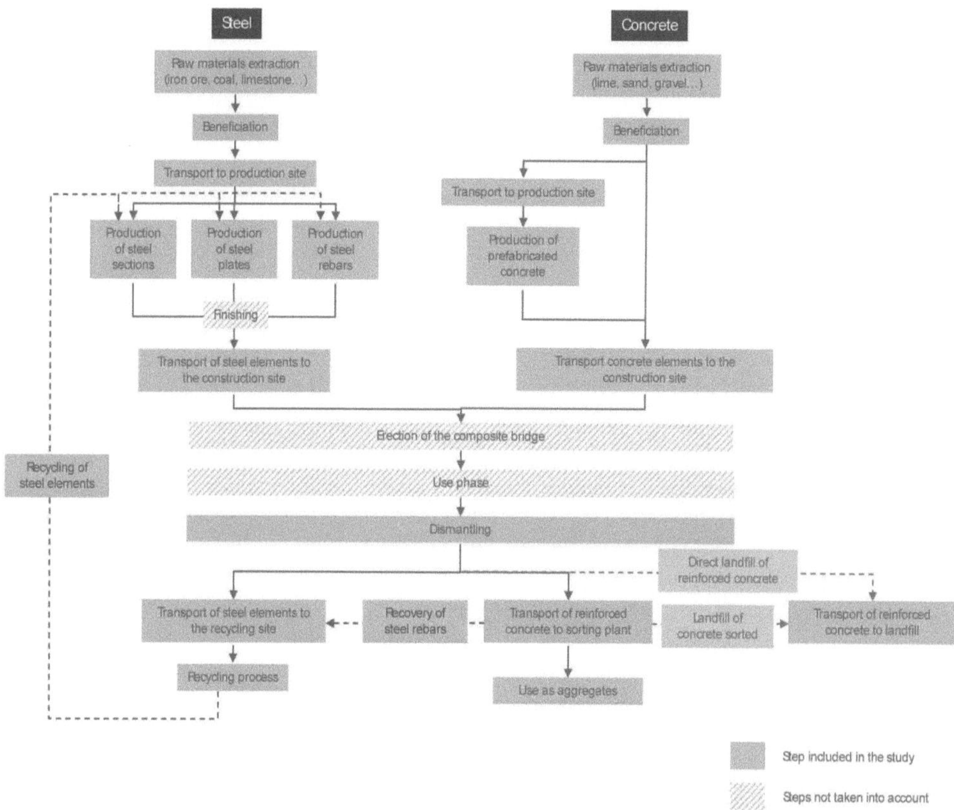

Figure 5:   Life cycle of the bridge

However, some stages of the life cycle are excluded from the study: the finishing steps of steel elements, the erection of the bridge and its use phase. These stages are shown in Fig. 5 by a different colour pattern.

The finishing of steel elements consists in cutting and welding. A previous study has shown that this step is negligible in an LCA of steel products because it is very small [4] and has therefore not been included in this evaluation. Similarly, we assumed that the erection phase was also fairly small and therefore was neglected. Finally the use phase of the bridges is equivalent between the two solutions and is therefore not taken into account.

4.   Life Cycle Inventory

4.1   General

The Life Cycle Inventory (LCI) collects the mass and energy balances for each process of the life cycle. LCIs for unit processes are listed in Tab. 2.

4.2   LCI of steel elements

The LCI of steel products are compiled by WorldSteel, from data collected on steel sites between 2005 and 2007 [5]. Among the 14 LCI of steel products of WorldSteel, four have been used in this LCA: "sections" for beams, "rebars" for shear studs and concrete reinforcement, and "plates" for end plates.

During the dismantling of the composite bridge, steel sections, end plates and shear studs are easily separated. Reinforcing steel which is physically linked to concrete is more difficult to recover. The structural steel is directly transported from the dismantling site to the recycling site, while reinforcing steel, embedded in concrete, is partially transported to a sorting plant, separated from concrete and transported to the recycling site while the remainder is sent to a landfill facility, cf. Fig. 6.

As a consequence, the recycling rate for sections, plates and studs is 99%, which corresponds to the findings of the European study on steel construction products [6] and 65% for reinforcing steel, in line with the statistics recorded by the Steel Recycling Institute [7] and the information gathered by [8].

Recycling of steel products avoids the production of virgin steel. The methodology used by WorldSteel to integrate this benefit is the multi step recycling method [9], which is implemented in a practical way by calculating an avoided impact that provides a credit to the steel elements recycled at the end-of-life of the bridge.

4.3   LCI of concrete

The cement content required for this type of application is 320 kg/m^3 of concrete [10]. We have used the unit process "concrete, normal, at plant" of the Ecoinvent database [11] for the production of concrete with 300 kg/m^3 of cement and extrapolated the results proportionally for 320 kg/m^3.

At the end of life, concrete can either be land filled, or crushed and used to replace aggregates [12]. After discussion with internal experts the following scenario has been chosen for the EOL of reinforced concrete (schematized in "Fig. 6"):

- 35% of reinforced concrete is directly sent to landfill – thus embedded rebars are also 100% land filled;
- 65% of reinforced concrete is sorted: it is crushed to separate rebars from concrete, that steel being 100% recycled;
- For concrete, it was considered that after the sorting plant, 15% of the concrete is used as aggregates and 85% is land filled.

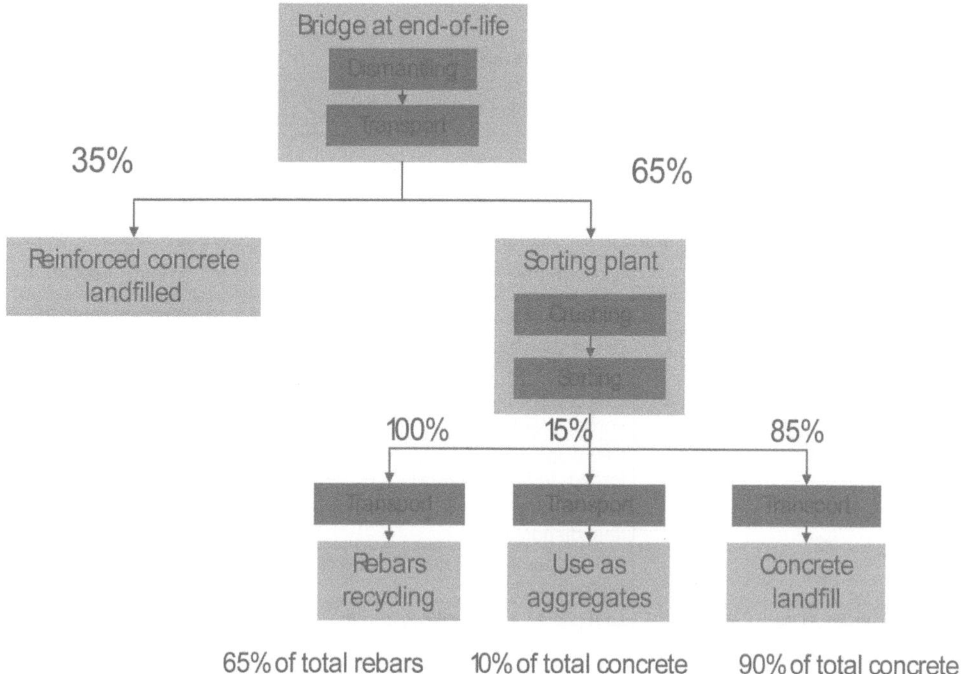

Figure 6:   End of life scenario – reinforced concrete

Use of EOL concrete avoids the consumption of aggregates. This environmental benefit is taken into account in the evaluation by providing a credit to the EOL concrete that is reused.

For concrete, LCIs corresponding to each EOL process are provided by the Ecoinvent database, as shown in Tab. 2.

4.4   *LCI of additional processes*

Steel and concrete elements are assumed to be transported by truck only, either regular truck for steel elements and prefabricated concrete, or mixer truck for ready-mixed concrete. The consumption linked to transportation is calculated by taking into account partial load and empty return trips [13]. Expert opinions were collected to estimate transport distances. Steel elements are supposed to be transported on an average distance of 1000km, and prefabricated concrete on 500km. Concerning on-site concrete, a transport distance of 50 km for the ready-mixed concrete is assumed.

Other information (fuel production, emissions linked to transportation, etc.) is provided by the models of the consulting group PE International [14].

Table 2:   List of data and sources for unit processes

Process	LCI associated	Source
Production and recycling of steel beams	EU: Sections	WorldSteel 2010
Production and recycling of steel studs	GLO: Studs – 99%	WorldSteel 2010
Production and recycling of steel end plates	EU: Plates– 99%	WorldSteel 2010
Production and recycling of reinforcement	GLO: Rebar– 65%	WorldSteel 2010
Concrete production	CH: concrete, normal, at plant	Ecoinvent 2011
Separation of concrete and reinforcement	CH: disposal, building, concrete, not rein-forced, to sorting plant	Ecoinvent 2011
	CH: disposal, building, reinforcement steel, to sorting plant	Ecoinvent 2011
Landfill	CH: disposal, building, concrete, not rein-forced, to final disposal	Ecoinvent 2011
	CH: disposal, building, reinforcement steel, to final disposal	Ecoinvent 2011
	CH: disposal, concrete, 5% water, to inert material landfill	Ecoinvent 2011
Use as aggregates	CH: gravel, unspecified, at mine	Ecoinvent 2011

5.   Life Cycle Impact Assessment

5.1   *Calculation of environmental burdens*

The Life Cycle Impact Assessment stage evaluates the potential environmental impacts of a product's life cycle by associating the LCI results (mass and energy balances) with the models of impacts, such as those listed in Tab. 1.

For example, let us describe the calculation method for the global warming indicator. The Global Warming Potential (GWP) is defined by the United Nations Framework Convention on Climate Change as "an index representing the combined effect of the differing times greenhouse gases remain in the atmosphere and their relative effectiveness in absorbing outgoing infrared radiation" [15]. This indicator was constructed by the Intergovernmental Panel for Climate Change. These climate experts have analyzed the radiative forcing of greenhouse gases and deduced a conversion factor for each greenhouse gas to an equivalent emission of carbon dioxide. For example, over 100 years, 1 kg of methane emitted will have the same radiative forcing as the emission of 25 kg of carbon dioxide.

5.2   *Comparison of environmental profiles*

In this paper,"environmental profile" refers to the environmental burdens generated during the life cycle of the bridges and characterized by the eleven indicators of Tab. 1.

Fig. 7 shows a comparison of the environmental profiles of the two bridges. To facilitate the reading, the indicators have been normalised to the burden of the concrete bridge. Abbreviations used in Fig. 7 are defined in Tab. 1.

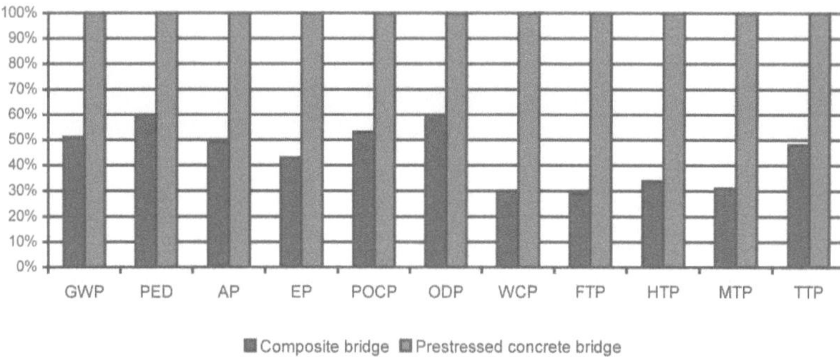

Figure 7:   Comparison of the environmental profiles of the two bridges

The environmental impacts of the composite bridge are lower than those of the pre-stressed concrete bridge for all indicators. The difference varies from 40% (primary energy demand, PED and ozone depletion potential, ODP) to 70% (for water consumption, WCP and freshwater, HTP, human, HTP, and marine toxicity, MTP).

This result is consistent with the bill of quantities shown in Fig. 4, the composite bridge being three times lighter than the prestressed concrete bridge.

5.3   *Contribution analysis*

In order to better understand the results presented in Fig. 7, the stages of the life cycles have been analysed to identify their hot spots. To this aim, we focus here on the results for the global warming indicator and show the contribution of each phase of the life cycle: production, end-of-life and overall life cycle impacts. As transportation is not differentiated among phases, it is added to the overall impact without further detail.

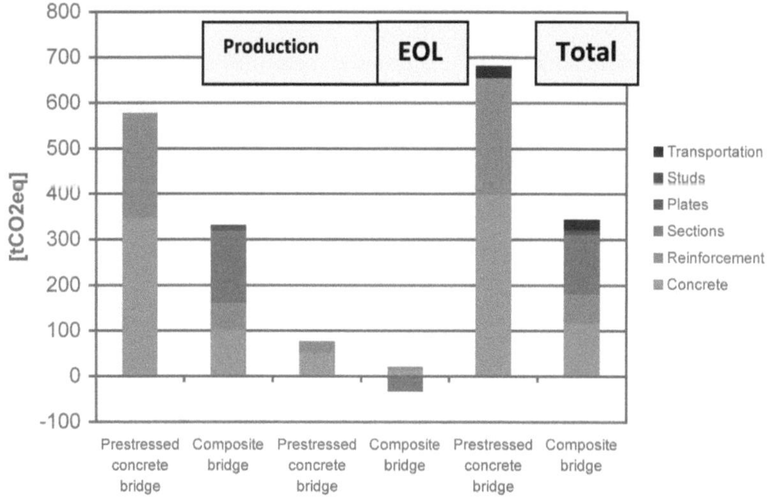

Figure 8:   Global Warming Potential of both bridges, distributed amongst production, transportation and end-of-life

Production phase

The GWP of the production phase is dominated by concrete production for case A, and by steel production for case B, which respectively accounts for 60% and 69% of the impacts. Comparatively, the composite bridge production (330 tCO2eq) generates 43% less GWP than the prestressed concrete (580 tCO2eq).

End-of-life phase

As explained in 4.2 and 4.3, a credit is given to materials that are reused or recycled at their end-of-life.

For the composite bridge, recycling of steel avoids 32 tCO2eq and reuse of concrete 0.2 tCO2eq. This is compensated by the impacts linked to the dismantling of the bridges, the treatment before reuse (separation of concrete and rebars) and the landfill (22 tCO2eq). Altogether, the EOL phase of the composite bridge has a negative value (-10 tCO2eq) and thus slightly decreases the overall life cycle GWP of the bridge.

On the other hand, for the prestressed concrete bridge, the GWP of the dismantling, separation of concrete and steel and landfill (78 tCO2eq) is larger than the credit provided by recycling (1 tCO2eq). As a result, the EOL phase of the prestressed concrete bridge has a positive value (77 tCO2eq) and thus increases the overall life cycle GWP.

These results are sensitive to a change in EOL assumptions; therefore another scenario was tested to evaluate the effect of a 100% recycling of reinforcing steel and concrete. In this scenario, all reinforced concrete needs to be crushed. In that case, the recycling compensates the burden of sorting only because of the recycling of steel while the benefit of recycling concrete is not high enough.

Altogether, for the two bridges, the EOL phase has a much lower contribution to the life cycle GWP than the production phase.

Life cycle results

The production phase (from extraction of raw materials to semi-finished products) is the main contributor to the GWP. This is also true for seven of the other indicators, except for ozone depletion (ODP), acidification (AP), human (HTP), marine (MTP) and freshwater toxicity (FTP) where the EOL phase contributes much more to the overall results. This is related to the treatments in the sorting plant which has a large effect on these particular indicators.

Finally, the contribution of transportation of concrete and steel elements represents a rather low share of the overall GWP results, respectively 4% and 7% for the prestressed concrete bridge and the composite bridge.

6.   Conclusions

There is a direct connection between the mass of the bridge as shown in the bill of quantities and the environmental burdens. The heavier concrete bridge has a significantly higher burden.

The production phase is more important than the end-of-life phase and the transportation of materials for most indicators. In addition, the end-of-life phase is sensitive to the fate of materials, whether they are recycled, reused, or land filled. The avoided impact by the EOL of reinforced concrete being smaller than that of the treatment necessary to crush it, the EOL of reinforced concrete downgrades the overall impact of bridges. On the contrary, the other steel elements (sections, plates and studs) avoid environmental impacts and improve globally the profile of the bridges.

In the case of the composite bridge, recycling of steel reduces the emissions by 10% (32 tCO_2eq) savings which are equivalent to 246.000 km driven by a car emitting 130 gCO_2/km [16].

The conclusion of this work shows clearly that the composite bridge has significantly smaller environmental burdens than the other bridge: this directly related to the use of steel in the first bridge. Moreover, the difference is very large: 40% to 70% depending on the indicator.

Another interesting result of this work is that recycling does not necessarily reduce the environmental burdens as the cost of recycling may be higher than the benefits it brings. A material like steel, the recycling of which avoids using virgin iron ore, should thus always be recycled – and indeed it is in general recycled to a very high level (more than 80%). A material like concrete, which is recycled as an aggregate, i.e. a low energy low greenhouse gas material, generates more burdens by recycling than by land filling. This is true of primary energy use and CO2 emissions, but is even more significantly true of ODP, AP, HTP, FTP and MTP.

When designing a bridge, it is essential to carry out an LCA and to use the results to improve on the design. Software tools such as AMECO [17], which calculates a simplified LCA of buildings or bridges focussing on energy and greenhouse gases, help to achieve this.

References

[1]   ISO 14040-44: "Environmental management – Life cycle assessment – Principles and framework" and "Environmental management – Life cycle assessment – Requirements and guidelines"; 2006

[2]   PE INTERNATIONAL, "Critical review report of the method for the environmental evaluation of bearing structures made of steel and concrete", 2010

[3]   J.-B. Guinée, "Handbook on Life Cycle Assessment", P GUINEE, 2001

[4]   A.-L. Hettinger, F. Labory, Y. Conan, "Environmental assessment of building structures made of steel or concrete", 2010

[5]   WORLDSTEEL, www.worldsteel.org, 2010

[6]   ECSC Final Report, "Life-cycle assessment (LCA) for steel construction", 2002

[7]   SRI, "Steel Recycling Institute", www.recycle-steel.org, 2008

[8]   M. Sansom, J. Meijer, "Life-cycle assessment for steel construction", SANSOM, 2002

[9]   BIRAT, "The value of recycling to society", 2006

[10] ZEMENT, "Bauberatung Zement – Expositionsklassen von Beton und besondere Betoneigenschaften", 2004

[11] D. Kellenberger, H-J Althaus, T. Künninger, "Concrete Products and Processes – Ecoinvent", ECOINVENT, 2007

[12] Jeannette Sjunnesson, "Life Cycle Assessment of Concrete" – Master thesis, LUND University, Department of Technology and Society, September 2005

[13] INRETS, Institut National de Recherche sur les Transports et leur Sécurité

[14] PE International – GaBi databases, 2010

[15] IPCC, http://unfccc.int/, 2010

[16]   Ademe report, "Car Labelling", http://www2.ademe.fr/, 2011

[17]   AMECO, "Software on Life Cycle Assessment of bearing structure", www.arcelormittal/sections, 2010

Design of composite dowels as shear connectors according to the German technical approval

Markus Feldmann[1], Daniel Pak[1], Maik Kopp[1], Nicole Schillo[1], Josef Hegger[2] & Joerg Gallwoszus[2]

Keywords: Composite dowels, composite beams.

Abstract: Composite dowels are known as powerful shear connectors in steel-concrete-composite girders. More and more they are used in practice especially for prefabricated composite bridges. Advantages over headed studs are in particular the increased strength, the sufficient deformation capacity even in high strength concrete and the simple application in steel sections without upper flange. However, missing provisions in standards for composite dowels with the economic clothoid and puzzle strip have led to retentions of clients and delays in the approval process. Hence, the aim of the recently finished German research project P804 [15] founded by FOSTA- Research Association for Steel Application was to solve open questions concerning these innovative shear connectors and to prepare a general technical approval available for any design office and construction company. In this paper design concepts for ultimate limit state and fatigue limit state, structural design principles and instructions for production and construction are presented and background information are given.

1. Introduction

Composite dowels are shear connectors for composite beams, which consist of openings in steel plates, casted with concrete. They are either made of steel plates welded on the upper flange of the steel beam or are fabricated directly out of the web of steel beams. Main advantages, compared to headed studs, are a higher bearing capacity and a sufficient deformation capacity even in high strength concrete to be classified as ductile shear connectors acc. to EN 199411. Furthermore, composite dowels are particularly suitable and economic for composite sections made of steel sections without upper flange, as steel parts next to the neutral axis are reduced ("Fig. 1 a"). However, they have been successfully applied for *VFT-Rail*® girders as well, where the compression zone is additionally reinforced by external reinforcement, consisting of composite dowels. Another economic application area is the arrangement of composite dowels in concrete tee-beams as external reinforcement ("Fig. 1 b"). In "Fig.2" headed studs and composite dowels are compared in view of longitudinal shear capacity as well as composite bending capacity and bending stiffness.

---

1   Institute of Steel Construction, RWTH Aachen University (Germany)
2   Institute of Structural Concrete, RWTH Aachen University (Germany)

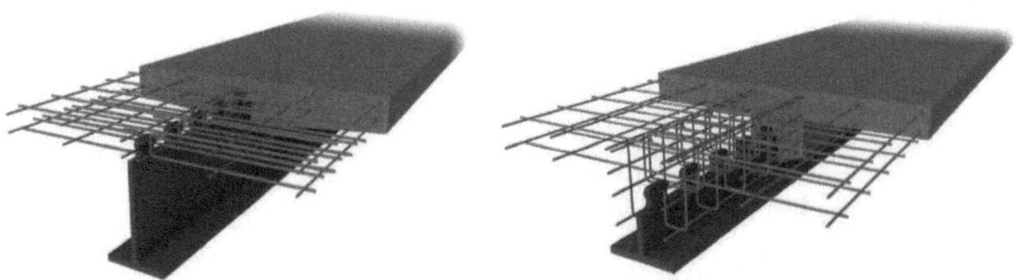

Figure 1: Application examples for composite dowels [3]

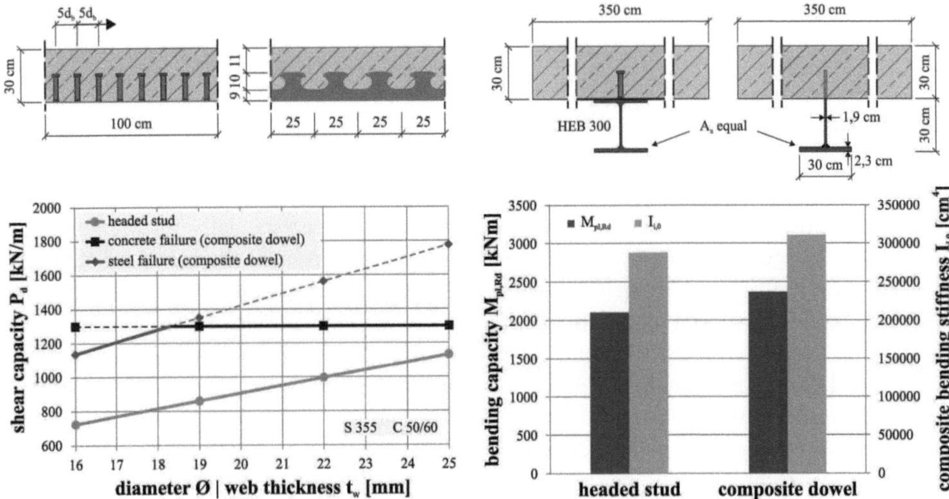

Figure 2: Comparison of longitudinal shear capacity between headed studs and composite dowels (left), comparison of moment capacity and stiffness between conventional composite section with headed studs and upper flange less steel section and composite dowels (right)

Up to now, the lack of technical rules for composite dowels led to delays in the approval process and to retention of clients. However, due to economic and technical advantages in more and more often composite dowels are used for road and railway bridges in Germany with approvals in the individual case. For example, the specific prefabricated composite bridge type *VFT-Rail®* with a composite dowel in clothoid strip was approved by the German railway authority (EBA) and applied e. g. in the railway bridge Simmerbach, Germany ("Fig. 3").

The starting point in the development of composite dowels can be traced back to research of Andrä and Leonhardt, which led to the development of the Perfobond strip [2], [11]. At the same time Bode developed the Kombidübel [1], [6]. Both design concepts are based on the mechanical model for shear failure of the concrete dowel. In the following years important knowledge about the bearing behavior of composite dowels was gained at the University of German Armed Forces [12], [13]. From this, mechanical models for the exceedance of the partial area pressure of concrete in the opening and concrete pry-out for composite dowels next to the concrete surface were obtained. The development of composite dowels with puzzle strip were pushed by research projects at RWTH Aachen University [8]

[10] [14] and by HOCHTIEF [9]. They led to a further development of the pry-out model and new models for steel failure. Seidl [16] developed design models for distance effects on concrete pry-out and vertical splitting of composite dowels in concrete tee-beams. The beginning of the clothoid strip – an optimization for fatigue loading – can be found in [4], [5]. This strip was finally used for the prefabricated composite bridge type *VFT-Rail®*, which was approved by the German railway authority (EBA) [7].

Figure 3: Bridges with composite dowels: road bridge in Pöcking (left) and Simmerbach (right) (photos: SSF Ingenieure AG, Munich)

2. Scope of the general technical approval

The new German general technical approval Z-26.4-56 [3] regulates composite dowels in clothoid (CL) and puzzle (PZ) strip ("Fig. 4"). The geometry can be scaled in dependence of the distance of the openings e_x between $150 \leq e_x \leq 500$ mm (notations see "Fig. 5"). The lower bound ensures a sufficient shear area for a ductile failure mode in the failure mode concrete shearing. The upper bound limits the maximum distance between composite dowels to prevent an unacceptable curtailment of the dowel capacity. The plate thickness can be varied between $6 \leq t_w \leq 60$ mm with a ratio of thickness to height between $0.08 \leq t_w / h_D \leq 0.5$. However, in the design formulas a thickness of up to 40 mm only is allowed to be considered. The minimum perpendicular distance of two steel plates is defined as 120 mm to ensure a sufficient installation of the reinforcement in between. For composite dowels structural steel in grade S235, S355 and S460 acc. to EN 10025 can be applied.

A minimum distance of 20 mm between concrete surface and top edge, respectively base, of the composite dowel has to be kept ("Fig. 5"). The distance from the opening to the concrete edge has to be in longitudinal direction more than 2.5 times the concrete pry-out cone h_{po} and in perpendicular direction more than 5.0 times of h_{po}. This ensures the full development of the concrete pry-out cone, if pry-out failure occurs. The minimum distance in perpendicular direction can be neglected, if in beam type sections the concrete reaches to the steel flange and confinement stirrups are installed (see below). This prevents pry-out at the lower concrete surface. A minimum width of beam type sections of 250 mm is required. For the composite section, concrete strengthC20/25 to C60/75 can be chosen (equal to the range given in EN 199411).

Composite dowels can be used under sagging and hogging moment for static as well as fatigue loading. However, structural members with centric tension forces perpendicular to the composite dowel under fatigue loading are not covered by the general technical approval as in this case. Unacceptable deterioration of the shear connection may occur.

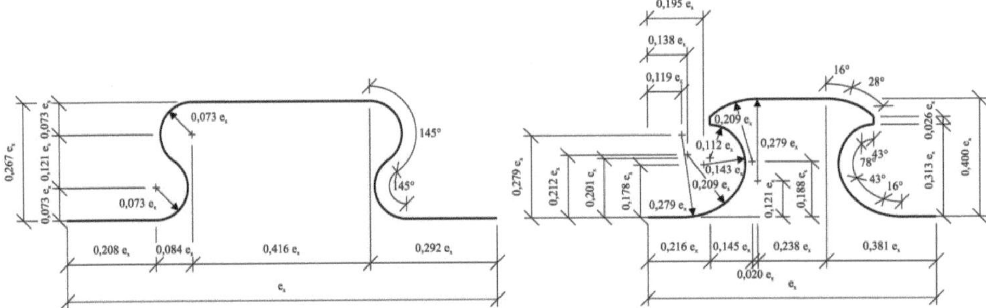

Figure 4:   Definition of the composite dowel geometry puzzle strip (left) and clothoid shape (right)

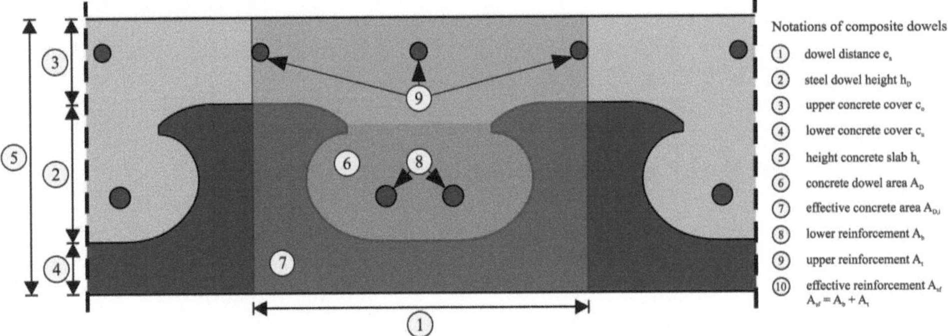

Figure 5:   Notations of composite dowels

3.   Design Concept

The design of composite beams with composite dowels is carried out in accordance with EN 1992, EN 1993 and EN 1994. The German general technical approval only regulates parts, which are not dealt with or which are different to the given European standards.

Besides the high bearing capacity, the major innovation leap of these composite dowels is:

- applicability in composite sections with steel beams without upper flange with equally spaced and partial shear connection;
- the ductility criterion according to EN 1994 is met to utilize plastic redistribution (for static loading only);
- a complete and consistent design concept for static and fatigue loading is provided for practical application.

4. Longitudinal shear capacity (static loading)

Possible failure modes of composite dowels subjected to static loading are concrete shearing,

- (i) concrete pry-out and
- (ii) steel failure.

The characteristic longitudinal shear capacity P_{rk} is determined as the minimum capacity of the aforementioned failure modes. The design value is calculated as the characteristic bearing capacity divided by a partial safety factor of $g_v = 1.25$.

The design formulas for these three failure modes are derived from existing, modified or new developed mechanical models, backed by a statistical evaluation of test results. The quality of different design concepts is compared by the ratio between experimental and theoretical values of a huge test data base [15]. Criteria are the shift of mean value and the coefficient of variation. The mechanical models with the highest quality are used to derive design formulas by the statistical evaluation procedure according to EN 1990 Annex D.

(i) Concrete shearing

In particular for small openings and large steel plate thicknesses the dominating failure mode is double shearing of the concrete dowel. Therefore, the main parameters for the bearing capacity are the shear area of the concrete dowel and the shear strength of the concrete. Furthermore, the bearing capacity is affected by the transversal reinforcement within the opening due to an additional doweling. In large openings the two shear areas merge together, which is considered by a geometry depended reduction factor η_D ($\eta_{D,CL} = 3 - e_x / 180$, $\eta_{D,PZ} = 2 - e_x / 400$). For sufficient large openings, which are guaranteed by the application range of the general technical approval, concrete shearing is a ductile failure mode.

$$P_{sh,k} = \eta_D \cdot e_x^2 \cdot \sqrt{f_{ck}} \cdot (1 + \rho_D) \qquad \text{in [N]} \tag{1}$$

with

$$\rho_D = \frac{E_s \cdot A_b}{E_{cm} \cdot A_D}$$

$$A_{D,CL} = 0{,}20 \cdot e_x^2 \quad \text{and} \quad A_{D,PZ} = 0{,}13 \cdot e_x^2$$

(ii) Concrete pry-out

Low distances between concrete dowel and concrete surface (top or bottom concrete cover) may provoke a failure mode, which is similar to the concrete pry-out of anchors subjected to shear forces. The hydrostatic pressure condition in the load introduction zone generates transversal tension forces, which lead – for insufficient concrete covers – to a cone-shape concrete pry-out ("Fig. 6"). The pry-out can occur for top or bottom concrete cover depending on the position of the concrete dowel. The result is a loss of hydrostatic pressure condition and therefore a reduction of concrete strength, which causes a secondary concrete pressure failure. This failure mode is ductile.

$$P_{po,k} = 90 \cdot h_{po}^{1,5} \cdot \sqrt{f_{ck}} \cdot \left(1 + \rho_{D,i}\right) \quad \text{in [N]} \tag{2}$$

$$h_{po} = \min\left(c_o + 0,07 \cdot e_x ; c_u + 0,13 \cdot e_x\right)$$

$$\rho_{D,i} = \frac{E_s \cdot A_{sf}}{E_{cm} \cdot A_{D,i}}$$

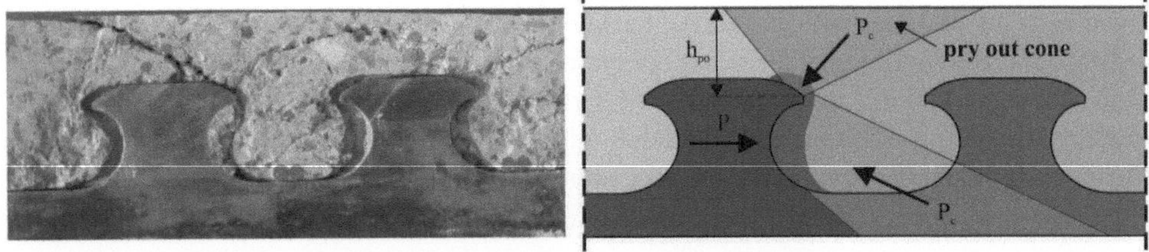

Figure 6: Failure mode pry-out: test specimen (left) and schematic illustration (right)

For composite dowels, where the distance in longitudinal direction is below $e_x < 4,5 \ h_{po}$, the bearing capacity is reduced due to an overlapping of the concrete cones. In this case $P_{po,k}$ has to be reduced by χ_x.

$$\chi_x = \frac{e_x}{4,5 \cdot h_{po}} \tag{3}$$

The same effect occurs in the case of parallel arrangement of composite dowels with a distance smaller than $e_y < 9 \times h_{po}$. For this the following reduction factor is used:

$$\chi_y = \frac{1}{2}\left(\frac{e_y}{9 \cdot h_{po}} + 1\right) \le 1 \tag{4}$$

In plate type sections, stirrups ø8 mm have to be used to guarantee for a ductile behavior. A maximum spacing of $4.5 \cdot h_{po}$ and 300 mm has to be kept to ensure that at least two reinforcing bars are in each concrete cone ("Fig. 7").

The design against concrete pry-out can be neglected, if the concrete is covered by a steel flange and confinement stirrups are applied (e. g. beam type sections with external reinforcement).

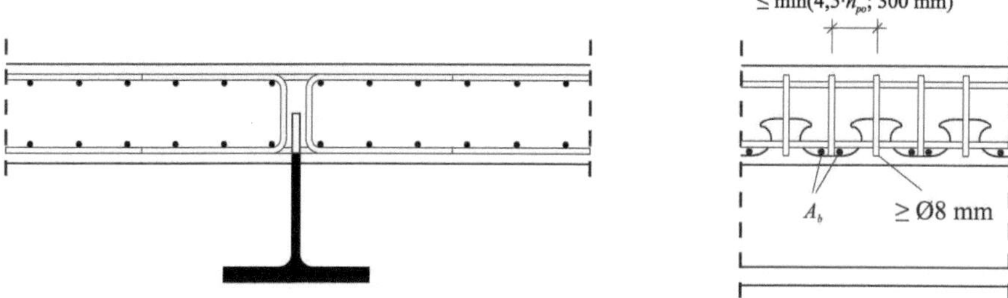

Figure 7:   Detailing of reinforcement in composite girder with RC-slab and composite dowels

(iii) Steel failure

For small plate thickness and low steel strength, steel failure of the steel dowel can occur. This failure mode is caused by a combined shear-bending mechanism, which leads to a horizontal crack in the steel plate. Due to the ductile behavior of structural steel, this failure mode goes along with large plastic deformations and is therefore ductile ("Fig. 8").

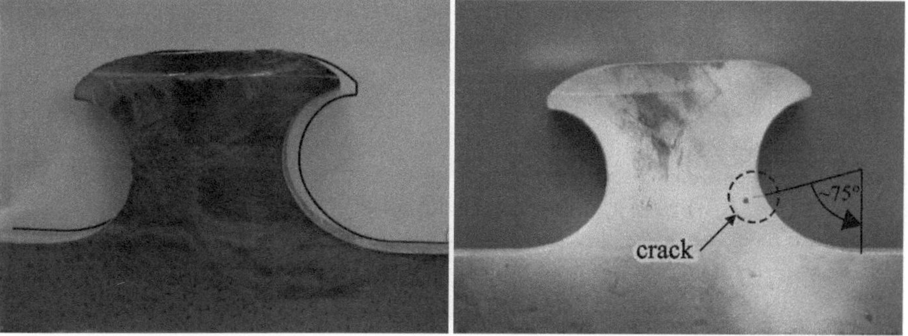

Figure 8:   Test specimen with steel failure: plastic deformation (left) and static crack (right)

The basis of the mechanical model is von Mises yielding in a critical section, where stresses due to the shear force P_2 and the bending moment due to $P_2 \cdot z_{p2}$ occur. The position of the critical section is an extremum problem of decreasing cantilever z_{p2} and decreasing load P_2 as well as a decreasing section area ("Fig. 9"). Based on this mechanical model the characteristic bearing capacity is expressed by:

$$P_{pl} = \frac{h_{eff}}{h_{eff} - h_{krit}} \cdot \frac{b_{krit}^2}{\sqrt{16 \cdot h_{s,krit}^2 + 3 \cdot b_{krit}^2}} \cdot t_w \cdot f_y$$

$$(5)$$

with $h_{krit}(\alpha)$, $h_{s,krit}(\alpha)$, $b_{krit}(\alpha)$, where $\alpha = f(\partial P/\partial \alpha \rightarrow \min)$

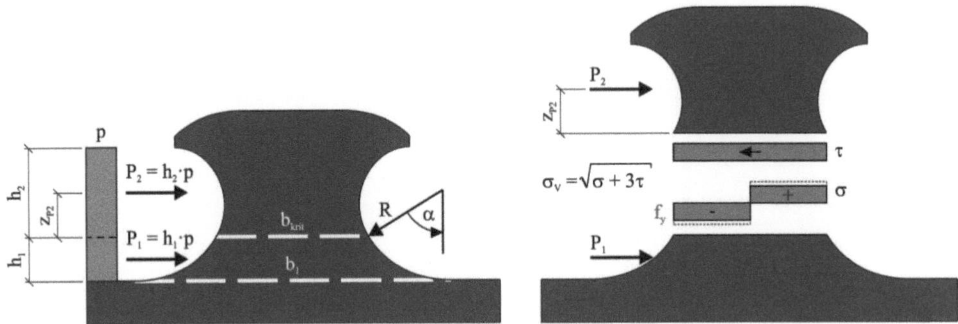

Figure 9:	Schematic illustration of failure mode steel failure

Using the specific geometrical parameters for the clothoid and puzzle shape the design formula can be rewritten as a function of e_x:

$$P_{pl,k} = 0,25 \cdot e_x \cdot t_w \cdot f_y \tag{6}$$

(iv) Additional verifications for beam-type sections

In beam-type sections with composite dowels as external reinforcement, failure of the concrete web can occur. For thin concrete webs splitting tensile forces can exceed the concrete tensile stress, which results in a horizontal crack at the height of the composite dowel ("Fig. 10"). This failure mode is non-ductile and has to be prevented by sufficient confinement stirrups. The required reinforcement $A_{s,conf}$ is determined in the line with models for pre-stressed concrete by a strut-and-tie model with an aperture angle of 33° (equ. 7). The confining stirrups have to enclose the concrete strut, which is guaranteed by a spacing smaller than (e_x; 300 mm) ("Fig. 11"). Furthermore, the stirrup has to reach at least $\Delta = 0,15 \cdot e_x$ below the dowel base. For a ductile behavior, a minimum of two stirrups (ø10 mm) for each opening is required.

$$A_{s,conf} = 0,3 \cdot P/f_{sd} \tag{7}$$

Additionally, shear forces in the concrete web have to be verified according to EN 199211. In this design check, the effective depth d_v starts from half the dowel height.

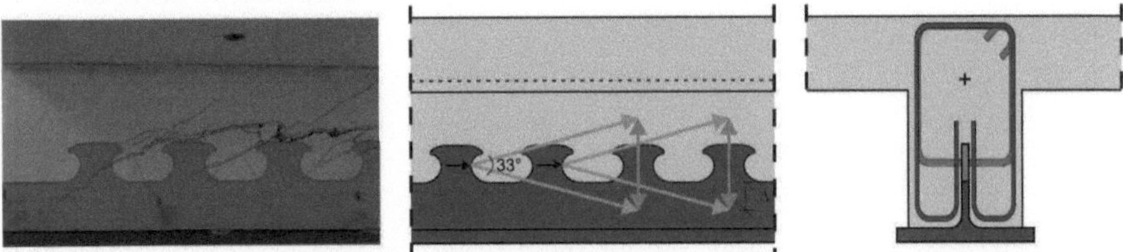

Figure 10:	Failure mode vertical splitting: test specimen (left), strut-and-tie model (middle) and required stirrups (right)

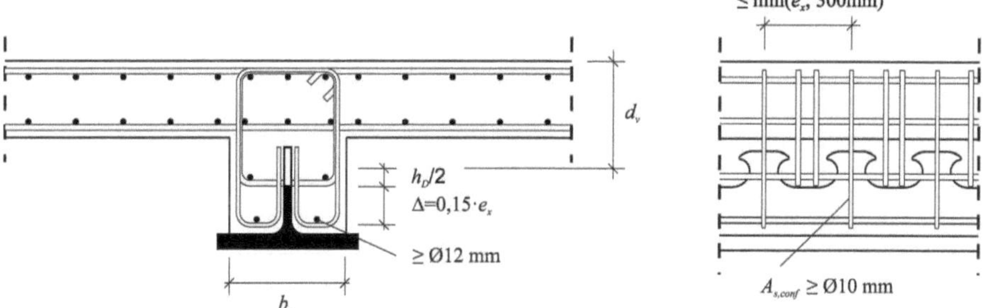

Figure 11:   Detailing of reinforcement in composite girder with RC-web and composite dowels

(v) Shear connection

The verification of the longitudinal shear connection has to be proofed in accordance with EN 199411. The number of required shear connectors equals the number of openings in the steel plate. The required transversal reinforcement is determined based on a 45° strut-and-tie model:

$$A_{s,conf} = 0{,}3 \cdot P / f_{sd} \tag{8}$$

Equal spacing of shear connectors is allowed for hogging and sagging moment, if the minimum shear connection degree acc. to EN 199411 is met. In contrast to EN 199411, equal spacing is also allowed for steel sections without upper flange, if the following requirements are met:

- the shear connection degree is $\eta \geq 0.5$;
- the span is $L \leq 18.0$ m;
- the plastic moment capacity of the composite section must be equal to or smaller than 10-times the plastic moment capacity of the steel profile;
- the curtailment of the dowel capacity must not incise the design longitudinal shear force more than 25 %.

Equally spacing for steel sections without upper flange extend the range of EN 199411 and opens new interesting application fields for composite dowels.

For fatigue loading the longitudinal shear force has to be determined by elastic theory and the curtailment of the dowel capacity must not incise the design longitudinal shear force. Reason for this is the prevention of unacceptable deterioration of the connection due to plastic force redistribution.

5.   Fatigue strength

The fatigue design has to be carried out according to the fatigue load model of appropriate standards. The fatigue design concept comprises

(i)   steel fatigue design,

(ii)   concrete fatigue design and

(iii) securing of a rigid shear joint.

Here, the interaction between the fatigue behavior of different components has to be considered ("Fig. 12"): e. g. degradation of the concrete dowel leads to decrease of the shear connection. This affects the stress distribution over the cross section and therefore the steel fatigue design.

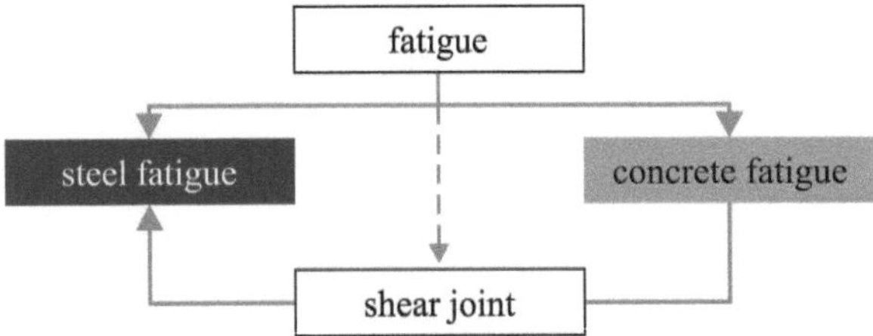

Figure 12:   Schematic illustration of the interaction of the fatigue verifications

(i) Steel fatigue

The steel fatigue design is based on the geometric stress approach. The stress amplitude at the hot spot is determined for the fatigue load model and compared with the material fatigue strength. The fatigue strength (resistance) is described by the fatigue strength curve of detail category 125 (machine gas cut edges having shallow and regular drag lines) or of detail category 140 (machine gas cut edges with subsequent dressing) in accordance with EN 199419. The geometric stress amplitude (action) is the sum of stresses due to longitudinal shear forces (local effects) and bending of the composite beam (global effects). Both parts are amplified by stress concentration factors depending on the geometry of the composite dowel. The nominal stresses are defined as longitudinal shear stress (local) and normal stress (global) at the dowel base.

$$\Delta\sigma = k_{f,L} \cdot \frac{\Delta V \cdot S_y}{I_y \cdot t_w} + k_{f,G} \cdot \left(\frac{\Delta N}{A} + \frac{\Delta M}{I_y} \cdot z_D \right) \tag{9}$$

with $\quad k_{f,L,CL} = 7{,}3 ; \quad k_{f,G,CL} = 1{,}5$ $\tag{10}$

$\qquad\quad k_{f,L,PZ} = 8{,}6 ; \quad k_{f,G,PZ} = 1{,}9$

The stress concentration factors kf,L (local) and kf,G (global) for the clothoid and puzzle shape are determined by finite element analysis and verified by strain measurements in cyclic push out and beam tests [15] ("Fig. 13"). The stress concentration factors are applicable for steel sections with a lower flange and concrete strength C20/25 and higher. To exclude low cycle fatigue, the geometric stress amplitude is limited to $2 \cdot fy$ and the upper geometric stress is limited to $1.3 \cdot fy$.

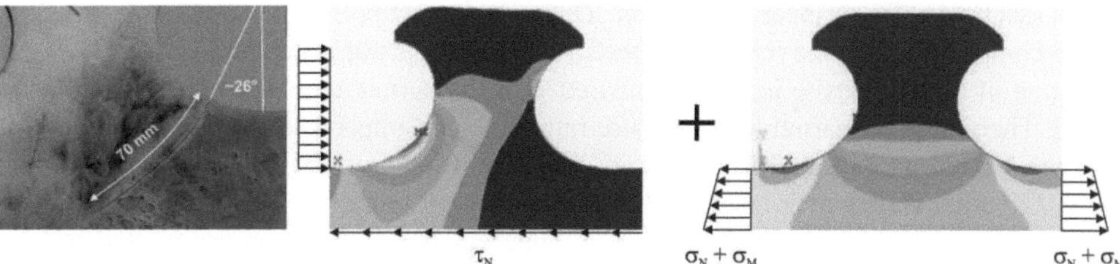

Figure 13: Steel fatigue failure: test specimen (left), FE-analysis of stresses due to local and global loading (middle and right)

In general tension stiffening has to be considered in the stress determination for regions where concrete cracking is expected. On the safe side the influence of tension stiffening (cracked section) can be neglected for the determination of stresses due to global bearing. For stresses due to longitudinal shear transfer the more unfavorable value from calculations with a cracked and an un-cracked section should be used. The design rule in EN 199411, which allows for the conservative determination of the fatigue stress in the composite joint with an un-cracked section, can be unsafe for composite dowels under hogging moment.

(ii) Concrete fatigue

Two types of concrete fatigue failure are known from tests:

- the loss of bearing capacity due to trickling of crushed concrete out of the composite joint;
- cyclic concrete pry-out of composite dowels with insufficient concrete cover subjected to high upper loads.

The first failure mode can be prevented by a limitation of the crack width to 0.15 mm in regions, where the composite dowel is in the concrete tension zone. This has to be considered for bending action in the concrete chord in longitudinal as well as perpendicular direction. Cyclic pry-out can be prevented by limitation of the upper load to 70 % of characteristic static bearing capacity according. to equation 1 and 2. Up to this load level the bearing behavior is mainly elastic.

(iii) Securing rigid shear connection

The basic requirement for the determination of stresses for the composite cross-section based on elastic theory is the assumption of a rigid shear connection between steel section and concrete section. However, results of cyclic beam tests show that this assumption is not justified a priori, which can lead to higher geometric stresses in the steel section (Fig. 14).

The loss of rigid shear connection is caused by a degradation of the concrete dowel due to cyclic loading. The consequence is a forceless slip before the composite dowel is activated (Fig. 14 bottom). This can be avoided by a limitation of the upper force. As criterion for a relevant degradation of the composite joint, the strain shift between steel and concrete section is used. Zero shift is equal to full shear connection; the maximum strain shift corresponds to a composite beam without shear connection. The threshold for an unacceptable degradation is assumed at 90 % shear connection. This criterion is applied to evaluate the condition of the composite joint in cyclic beam tests (acceptable or not acceptable). Afterwards the condition of the composite joint is correlated with the partial area pressure at the concrete dowel. Therefore, the partial area pressure ratio $E_{cd,3D}$ for upper and lower load from beam tests are plotted into a Goodman diagram, which is often used for concrete fatigue design ("Fig. 15" left). Tests with a residual shear connection of more than 90 % are located in the green area (acceptable), while tests with an unacceptable degradation are plotted outside. This behavior is shown in "Fig. 15" (middle and right). The strain distribution in test series 7 shows a relatively large strain discontinuity $\Delta\varepsilon$ compared to test series 8, where the loss of rigid shear connection was not that distinctive. The evaluation of the tests shows that the available concrete compressive strength under multi axial stress condition is up to 7.5 times higher than the uniaxial compressive strength. Based on the assumption that at least 10 % of the total loads are dead loads, the upper load is limited to 55 % of the available multi axial compression strength.

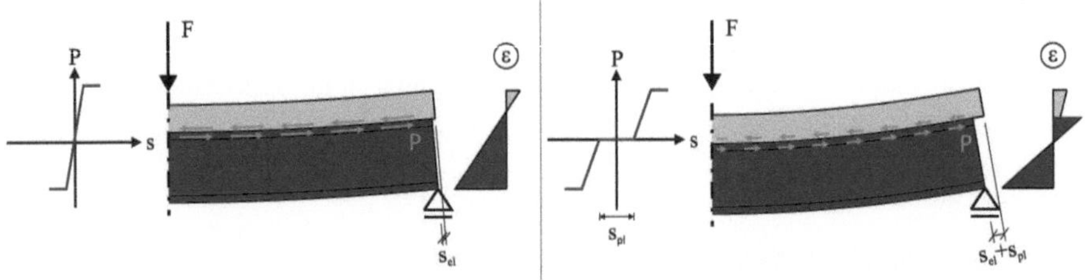

Figure 14:   Deformation behavior and strain distribution of composite girder with intact shear connection (left) and damaged shear connection (right)

$$P_{cyc} = 3{,}1 \cdot t_w \cdot h_D \cdot f_{ck} \tag{11}$$

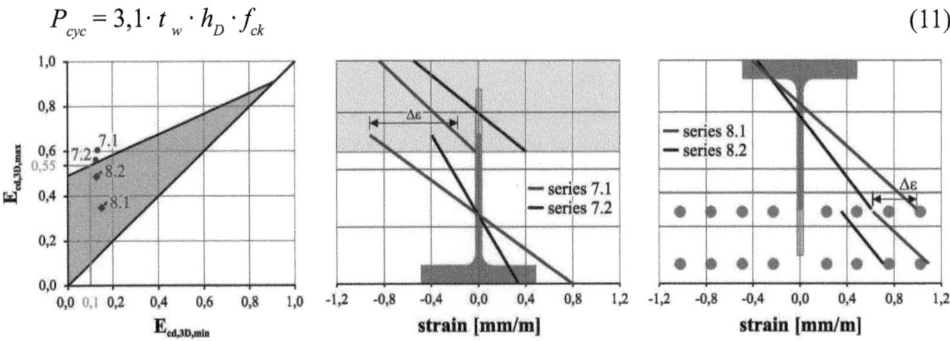

Figure 15:   Fatigue strength of cyclic loaded beam tests as a function of partial area pressure ratio $E_{cd,3D}$ for upper and lower load

6.   Fabrication and construction

Composite dowels have to be fabricated by gas cutting or a cutting-process, which is similar in terms of strength and fatigue. The nominal geometrical values are given in Fig. 4, where tolerances of +2/-4 mm are acceptable ("+" is an enlargement of the steel shape). The cutting quality has to be in accordance with EN 10901 and EN 10902 and depends on the execution class. For fatigue loading the quality has to meet the requirements for detail category 125,

140 respectively, according to EN 199319. Checking and documentation of the cutting quality is very important as inspection of the shear connector is not possible after casting. To prevent blowholes in the concrete next to the composite dowel, the maximum grain size is limited to 16 mm and the consistence of the fresh concrete should be soft to flowable.

7. Outlook

The general technical approval is a consistent continuation of the development of standardization of composite dowels: it started with approvals for single cases and should become a German steel guideline (DASt) in the future. For this, the presented design concept is a strong basis. Due to the given regulations uncertainties in design of composite dowels are eliminated. This enables an economic and timesaving application of these powerful shear connectors. Additional costs for approvals for single cases are omitted. For the future, it is planned to implement composite dowels in Eurocode 4 as alternative shear connector besides headed studs.

8. Notations

$P_{sh,k}$	[N]	Characteristic value of the concrete shearing resistance
$P_{po,k}$	[N]	Characteristic value of the concrete pry-out resistance
$P_{pl,k}$	[N]	Characteristic value of the steel resistance
e_x	[mm]	Distance of recesses in longitudinal direction
η_D	[-]	Reduction factor depend on the dowel geometry
η_D	[-]	Reinforcement ratio for transversal reinforcement
$\eta_{D,i}$	[-]	Effective reinforcement ratio for transversal reinforcement
E_s	[N/mm²]	Young's modulus of reinforcing steel
E_{cm}	[N/mm²]	Secant modulus of elasticity of concrete
A_b	[mm²]	Lower reinforcement in the range of A_D
A_t	[mm²]	Upper reinforcement in the range of e_x
A_D	[mm²]	Concrete dowel area
$A_{D,i}$	[mm²]	Effective concrete area ($ex \cdot hc$)
A_{sf}	[mm²]	Effective reinforcement ($A_b + A_t$)
c_o	[mm]	Thickness of upper concrete cover
c_u	[mm]	Thickness of lower concrete cover
h_{po}	[mm]	Pry out cone height
f_{ck}	[N/mm²]	Characteristic value of the cylinder compressive strength of concrete
f_y	[N/mm²]	Nominal value of the yield strength of structural steel
χ_x	[-]	Reduction factor for concrete pry out in longitudinal direction
χ_y	[-]	Reduction factor for concrete pry out in transversal direction

e_y	[mm]	Distance between parallel arrangement composite dowels
t_w	[mm]	Web thickness
$A_{s,conf}$	[mm²]	Required transversal reinforcement in composite girders with RC-web
f_{sd}	[N/mm²]	Characteristic value of the yield strength of reinforcing steel
$\Delta\sigma$	[N/mm²]	Stress range
$k_{f,L}$	[-]	Stress concentration factor for longitudinal shear force (local)
$k_{f,G}$	[-]	Stress concentration factor for bending and axial force (global)
ΔV	[N]	Design value of the shear force range
S_y	[mm³]	First moment of area of the effective composite section
I_y	[mm⁴]	Second moment of area of the effective composite section
ΔN	[N]	Design value of the axial force range
A	[mm²]	Cross-sectional area of structural steel
ΔM	[Nmm]	Design value of the bending moment range
z_D	[mm]	Lever arm
P_{cyc}	[N]	Maximum force for cyclic loading
h_c	[mm]	Concrete slab height

References

[1] Allgemeine bauaufsichtliche Zulassung, Kombi-Verdübelung, Z-26.4-39, Deutsches Institut für Bautechnik DIBt, Berlin, 2000.

[2] Allgemeine bauaufsichtliche Zulassung, Perfobond-Leiste, Z-26.4-38, Deutsches Institut für Bautechnik DIBt, Berlin, 2007.

[3] Allgemeine Bauaufsichtliche Zulassung, Verbunddübelleisten, Z-26.4-56, Deutsches Institut für Bautechnik DIBt, Berlin, 2013.

[4] Berthellemy, J., Lorenc, W., Mensinger, M., Rauscher, S, Seidl, G., „Zum Tragverhalten von Verbunddübeln – Teil 1: Tragverhalten unter statischer Belastung", Stahlbau 80, pp. 172–184, 2011.

[5] Berthellemy, J., Lorenc, W., Mensinger, M., Ndogmo, J., Seidl, G., „Zum Tragverhalten von Verbunddübeln – Teil 2: Ermüdungsverhalten", Stahlbau 80, pp. 256–267, 2011.

[6] Bode, H., Künzel, R., „Scherversuche zum Tragverhalten eines neuartigen Stahlverbundträgers mit schwalben-schwanzförmigen Stegausnehmungen als Verbundmittel", reserach report 2/88, University Kaiserslautern, 1988.

[7] Eisenbahnbundesamt, „Bescheid zur Bauartzulassung des Systems VFT-Rail auf der Grundlage des Bemessungskonzeptes MV-TR", Bonn, 2010.

[8] Feldmann, M., Hechler, O., Hegger, J., Rauscher, S., „Neue Untersuchungen zum Ermüdungsverhalten von Verbundträgern aus hochfesten Werkstoffen mit Kopfbolzendübeln und Puzzleleiste", Stahlbau 76, pp. 826–844, 2007.

[9] Gündel, M., Dürr, A., Hauke, B., Hechler, O.,"Zur Bemessung von Lochleisten als duktile Verbundmittel in Verbundträgern aus höherfesten Materialien", Stahlbau 78, pp. 916–924, 2009.

[10] Heinemeyer, S., Gallwoszus, J., Hegger, J., „Verbundträger mit Puzzleleisten und hochfesten Werkstoffen", Stahlbau 81, pp. 595-603, 2012.

[11] Leonhardt, F., Andrä, W., Andrä, H.-P., Harre, W., „Neues, vorteilhaftes Verbundmittel für Stahlverbund-Tragwerke mit hoher Dauerfestigkeit", Beton- und Stahlbetonbau 82, pp. 325-331, 1987.

[12]   Mangerig, I., Wagner, R., Burger, S., Wurzer, O., Zapfe, C., „Zum Einsatz von Betondübeln im Verbundbau (Teil 1) – Ruhende Beanspruchung", Stahlbau 80, pp. 885–893, 2011.

[13]   Mangerig, I., Wagner, R., Burger, S., Wurzer, O., Zapfe, C., „Zum Einsatz von Betondübeln im Verbundbau (Teil 2) – Nichtruhende Beanspruchung", Stahlbau 81, pp. 26–31., 2012.

[14]   Feldmann M., Hegger J., Hechler O., Rauscher S., „P621 – Untersuchungen zum Trag- und Verformungsverhalten von Verbundmitteln unter ruhender und nichtruhender Belastung bei Verwendung hochfester Werkstoffe," FOSTA – Forschungsvereinigung Stahlanwendung e. V., Düsseldorf, 2007.

[15]   Feldmann M., Hegger J., Gündel M., Kopp M., Gallwoszus J., Heinemeyer S., Seidl G., Hoyer O., „P804 – Neue Systeme für Stahlverbundbrücken – Verbundfertigteilträger aus hochfesten Werkstoffen und innovativen Verbundmitteln. VERBUND-DÜBEL-LEISTEN," FOSTA – Forschungsvereinigung Stahlanwendung e.V., Düsseldorf, not published yet.

[16]   Seidl, G., „Verhalten und Tragfähigkeit von Verbunddübeln in Stahlbetonverbundträgern", dissertation, Wroclaw University of Technology, 2009.

Field measurements at a composite bridge with composite dowels as shear connectors

Markus Feldmann[1], Daniel Pak[1], Maik Kopp[1] & Nicole Schillo[1]

Keywords: composite bridge, composite dowel, external reinforcement, hot-spot stress, in-situ measurement, railway bridge, rail support

Abstract: Within the demonstration project ECOBRIDGE, funded by the Research Fund for Coal and Steel (RFCS), a monitoring program has been developed and applied to survey a composite railway bridge. The stress distribution between concrete and specially designed composite dowels, which have a better fatigue resistance compared to conventional headed studs, has been investigated as well as the forces in several rail support points. The German railway operator DB Netz AG requested the survey to prove the applicability of state of the art calculation methods, which were not part of compulsory codes.

1. Pilot bridge Simmerbach

In the course of railway track 3511 between Bingen and Saarbrücken (Germany), two existing steel troughs with ballast bed, which had reached their life-span, have been replaced by two *VFT-Rail®* girders with a span of 17.75 m each ("Fig. 2"). The composite *VFT-Rail®* system is characterized by two characteristics: first, composite action between concrete and steel girders (which act as external reinforcement) is ensured by means of composite dowels. These are cut out of an I profile ("Fig. 1", right). Second, the rail support points are directly fastened to the composite girder ("Fig. 1", left).

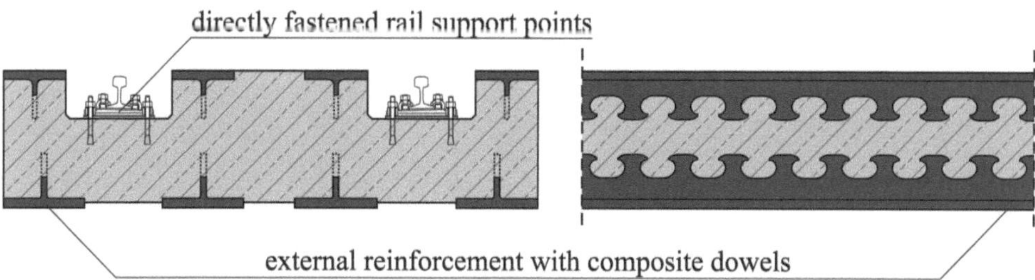

Figure 1:   Transverse cross section (left) *VFT-Rail®* with rail support points directly fastened to the superstructure and composite dowel and longitudinal cross section (right, depicting the clothoidal shape of the dowels).

---

1   Institute of Steel Construction, RWTH Aachen University (Germany)

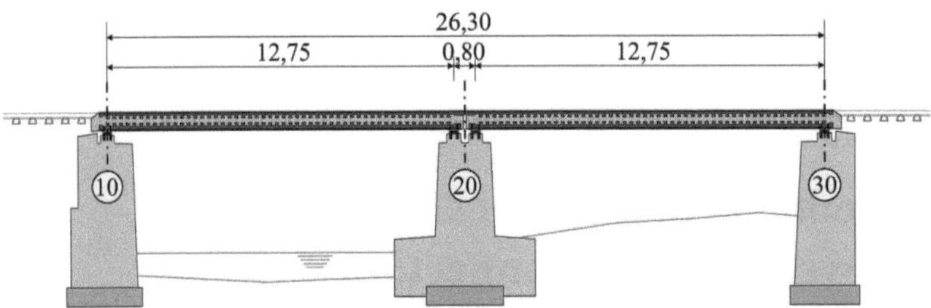

Figure 2:   Longitudinal section railway crossing over the river "Simmerbach" with two *VFT-Rail®* girders, placed on toughened sub-structure

A detailed description of the railway bridge is given in [1].

The rail operator made the following detailed long-term observation conditional as part of a special approval for the German market:

- measurement of strains in the steel girder caused by train crossings;
- measurement of forces acting on the rail support points;
- measurement of settlement of the first sleeper behind the bridge, responsible for an increase of tension and compression forces on the first four rail support points.

A detailed description of the measurements undertaken is given in this paper.

1.1   *Design and construction*

Both superstructures are designed as simply supported composite beams. Each single rail is placed in a channel to increase the effective height of the cross section ("Fig. 1", left). Four halved steel profiles are placed at the bottom side of the bridge taking tension. The upper part of the structure is reinforced with four halved steel profiles as well. This additional external reinforcement is needed to carry compression forces, as the active concrete cross section is reduced by the channels, giving space for the rails. Composite action between steel girders and concrete is realized by composite dowels (geometry MCL250/115, [2]). Shear reinforcement is placed in cut-outs between the steel dowels according to [3]. Due to concrete shrinkage, condition II builds up over the concrete's cross section. Therefore, for design purposes shear forces are assigned to the concrete, bending forces to the structural steel.

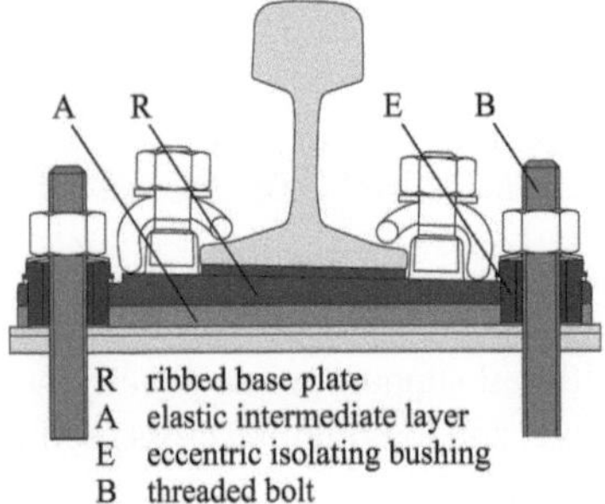

Figure 3:   Structure rail support point ECF [4]

The elastic, adjustable rail support system Krupp ECF ([4], 131-02) has been chosen to fasten the rails to the composite girder ("Fig. 3"). Each rail support point is anchored by means of two threaded bolts which are embedded in the precast concrete. Lining plates are used to compensate vertical tolerances. Horizontal tolerances up to +/- 10mm are compensated by means of eccentric insulating bushings.

The rail support points are placed in a regular pattern of a = 0.60 m. Next to the transition between superstructure and backfilling (axis 20 and axis 30), an additional point is used to reduce the compressive forces in this region ("Fig. 4").

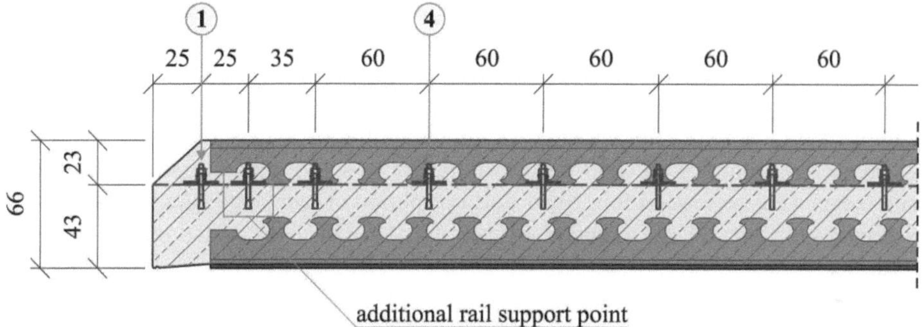

Figure 4:   Pattern of rail fasteners at the end of superstructure

The regular distance of 60 cm is continued on the ballast bed in front of the bridge. To ensure for small deformations under loading in that region, B90 sleepers are installed in front and behind the bridge structure. A possible maintenance replacement of the first four rail support points needs to be ensured, as the calculation of the tension forces acting on these fasteners is based on several assumptions such as time dependent ballast settlement in front of the bridge. Therefore the first four rail support points are fixed by means of stainless steel threaded bolts, screwed into friction welded stainless steel sleeves, anchored in the superstructure.

1.2   *Aim of field measurements*

The innovative aspects of the bridge structure are related to the composite dowel as well as the transition between ballast bed and directly fastened rail support points.

Therefore the following issues were investigated by means of field measurements:

- Confirmation of considerations underlying the design of the composite dowel, confirmation of correct application of rules based on stress concentration factors (section 4);
- Influence of settlement and deformation of ballast bed in front of the bridge on the forces acting on the rail support points (section 3).

Regarding the rail support points, threshold values were defined by the railway operator DB Netz AG which compliance had to be checked during monitoring. The applied monitoring concept was developed within the scope of the European research project ECOBRIDGE [5], co-financed by the Research Fund for Coal and Steel (RFCS) of the European Commission.

2.   Monitoring concept

Due to symmetric reasons, all measurement devices were applied at the *VFT-Rail*® girder between axis 20 and axis 30 of the bridge structure only ("Fig. 2"). For measurement of the strains in the composite dowel ("Fig. 1", right), strain gauges were applied and connected by wire during prefabrication of the elements in the workshop (section 4). Forces acting on the rail support points N° 1 and N° 4 were measured by means of two conventional ECF rail support points, equipped with special measurement devices (section 3). Furthermore relative settlement and lowering of the first B90 railway sleeper in front of the bridge was measured by means of two inductive displacement transducers. The temperature distribution over the cross section was recorded permanently by three internal temperature gauges. These measurements were completed by deformation measurements of the superstructure.

Continuous measurements (C) were performed in combination with temporary measurements (T) to allow for a low-frequency recording of long term effects as well as a high-frequency recording of short term loading due to single train crossings. Temporary measurements were performed four times during a period of two years (summer and winter measurements to cover a large temperature spectrum). In Tab. 1 the measurements are summarized.

Table 1:   Overview monitoring concept (type of measurement, type of data, N° of devices, sampling rate)

	type	data	N° of devices	sampling rate
strain in steel component	T	relative	72	1000 Hz
forces on rail support points	T / C	absolute	14	1000 Hz / -
settlement of railway sleeper	T / C	absolute	2	1000 Hz / -
vertical deformation of superstructure	T	relative	1	1000 Hz
temperature	C	absolute	3	-

For static measurements a train of known weight was used (diesel electric locomotive BR 132 "Ludmilla", load per axle = 204 kN, "Fig. 5"). In total three „stop-runs" were performed, whereas the locomotive came to a stop every 2 meters on the *VFT-Rail*® girder for at least one minute.

Figure 5:   Test train, class BR 132, Co'Co'

For dynamic measurements the same test train was used, whereas data was recorded during 12 train crossings at speeds of 20, 60, 80 and 100 km/h. These measurements allow for a quantification of the dynamic behavior of the structure [6] [7] [8]. Additional measurements were performed during train crossings of scheduled passenger trains ("RegionalExpress"), consisting of two class DBAG 612 railcars ("Fig. 6").

Figure 6:   Railcar class DBAG 612, 2'B'+B'2'

3.   Rail support points

The forces acting on the rail support points as well as the stresses in the rails at both ends of the superstructure were determined during the design process by means of a structural analysis (section 3.1). These calculations were based on national railway codes "Ril 804 Module 5404" [9] and "Ril 804 Module 5405" [10] which were not compulsory.

3.1   Determination of design values

The theoretic values of rail stresses and forces acting on the single rail support points were determined by means of a framework model. This model consisted of the superstructure (layer 1) and the two single rails (layer 2). These rails were continuing with a length of $L_D =$ $0.5 \cdot L_T + 40$ m $= 46.58$ m according to DIN Fachbericht 101, annex K.3.4 (4) [11] adjacent to the superstructure in both directions.

The rails were coupled to the superstructure by horizontal and vertical springs, representing the rail support points ("Fig. 4"). The vertical spring stiffness of the rail support point was governed by the properties of the elastomer bearing ("Fig. 7"), whereas the stiffness (22.5 kN/mm = 22,500 kN/m) was determined in advance by TU Munich (internal research report N° 1689a) for design purpose.

The horizontal creep resistance for the horizontal coupling springs was determined according to DIN Fachbericht 101, figure K.3 [11], as the rails are fixed directly to the superstructure. For the load case "temperature difference" the values for an unloaded track (c = 18,000 kN/m per single rail, yield value F = 9.0 kN/m) were used. For the load case "braking" the corresponding values for a loaded track (c = 36,000 kN/m per single rail, yield value F = 18.0 kN/m) were considered.

In the region of backfill the vertical spring stiffness was calculated by combining the spring stiffness of the rail support point and the spring stiffness of the elastic bedding (sleepers on ballast bed). For that case, a value of 0.10 N/mm² was chosen according to [12], which resulted in a total spring stiffness of 14,300 kN/m for each single vertical spring.

Using the design rules provided by [11], forces acting on the rail supports and stresses in the structure were calculated. These values were compared to the insitu measurements as shown in chapter 3.2.2.

The maximum stresses in the rail track at the abutments due to horizontal forces were determined to be 37.1 N/mm². This value by far did not exceed the limiting value of 92 N/mm² as given by DIN FB 101, annex K.3.6 [11].

The maximum stresses in the rails due to bending summed up to 170.6 N/mm².

Forces acting on rail support points due to temperature loading were:

- max $F_z =$ 16.40 kN (at 1st rail support point on superstructure due to temperature induced vaulting);

- min F_D = -4.60 kN (at 1st rail support point due to temperature induced flexion).

Forces acting on rail support point due to load scheme "Betriebszug type 5" were:

- max F_Z = 4.09 kN (at 4th rail support point);
- min F_D = -65.5 kN (at 1st rail support point).

Forces acting on rail support point due to load scheme "test train BR 132" were:

- max F_Z = 4.0 kN   (at 4th rail support point);
- min F_D = -36.3 kN (at 1. rail support point).

Alternatively a settlement of 3 mm was defined for the first springs on the backfill, which means that the springs did not act for settlements not exceeding 3 mm (realistic settlement). This proceeding resulted in higher forces acting on the rail support points for load scheme "test train BR 132":

- max F_Z = 6,79 kN (at 4th rail support point);
- min F_D = -51,90 kN (at 1st rail support point).

3.2   *Field measurements*

3.2.1   Test set-up and measurement

Wheel loads are transferred from the rail track to the superstructure by ECF rail support points at regular intervals ("Fig. 4"). The load transfer mechanism for vertical loads F had to be considered separately for compression and tension forces.

Compression forces are transferred from the rail through a ribbed base plate R and an elastic intermediate layer A into the superstructure ("Fig. 7").

The transfer of tension forces is realized by means of pre-stressed (stainless steel) threaded bolts B which are screwed into friction welded stainless steel sleeves, anchored in the superstructure ("Fig. 7").

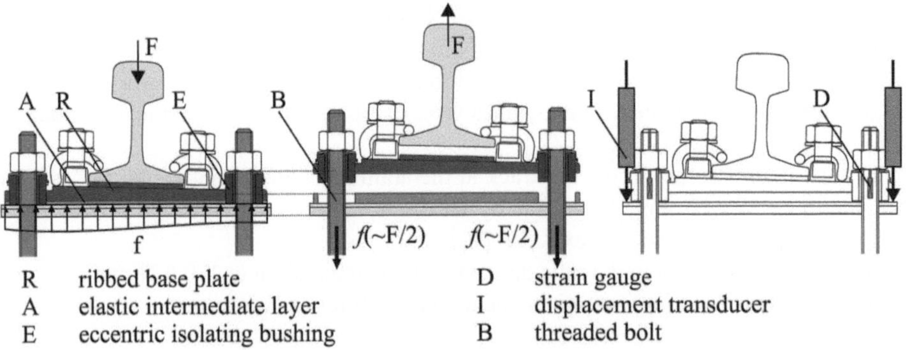

Figure 7:   Load bearing mechanism of rail support point ECF (principle, superelevated illustration), measurement equipment at rail support point ECF (principle)

As already mentioned, forces acting on the rail support point are not allowed to exceed limit values predefined by DB Netz AG. Within the scope of the bridge monitoring these vertical loads were measured separately for compression and tension forces, based on the corresponding load bearing mechanism.

In the case of an external tensile loading F acting on the rail support point, the tension force is transferred from the ribbed base plate R through eccentric insulating bushings with collar E into two pre-stressed stainless steel threaded bolts (measurement bolt) B, anchored in the superstructure. The determination of these tensile forces was realized by a direct measurement of normal strain in the measurement bolts for which special strain gauges are glued into the bolts ("Fig. 7", D and "Fig. 8").

The threaded bolts are not pre-stressed against the ribbed base plate R but against the eccentric insulating bushings E. Hence a compression loading of the rail support point did not lead to a measurable decrease of preload force in the measurement bolts, as the pressure hull (caused by pre-stressing of the bolt) built up in the eccentric insulating bushings E were not decreased by a lowering of the ribbed base plate R ("Fig. 7", "Fig. 11"). Therefore compression forces were determined by a measurement of the lowering of the ribbed base plate R combined with the known stiffness of the elastic intermediate layer. This approach is equivalent to the determination of vertical spring stiffness for design purposes as described in chapter 3.1. The actual deformation of the ribbed base plate R was gauged by four inductive displacement transducers relatively to the slabbed track ("Fig. 7", I and "Fig. 8").

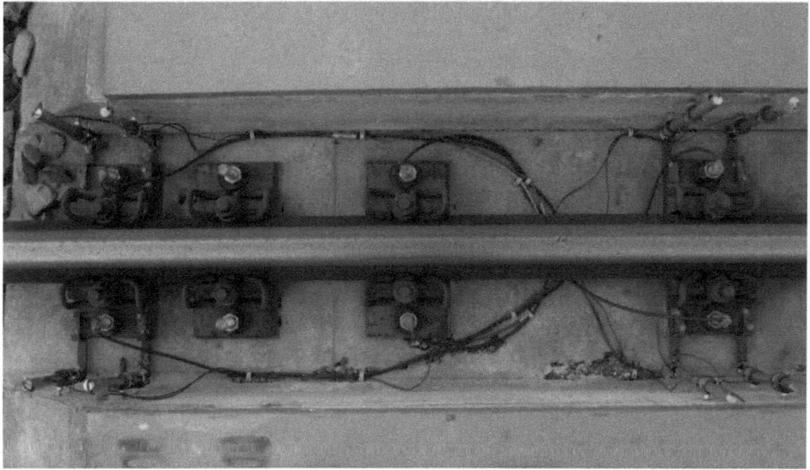

Figure 8:   Measurement devices for monitoring of forces acting on rail support points (1st, 2nd and 4th rail support point equipped)

In the first instance all measurement systems were calibrated at the Institute of Steel Construction of RWTH Aachen University in a phased process. Initially a linear calibration factor *ktension,1* was determined. This factor was used for the conversion of measured elastic strains into equivalent bolt forces. Based on the concept of measurement bolt calibrations already successfully applied in the former research project RFCS HISTWIN [13] [14], the measurement bolts were loaded displacement-controlled in a two-column testing machine up to a maximum load of F=50 kN. The strains measured during testing were plotted against the corresponding jack forces ("Fig. 9") to derive a linear calibration factor *ktension,1* for each individual bolt. These linear calibration factors are summarized in Fig. 9 for all measurement bolts, representing the mean of a series of two measurements.

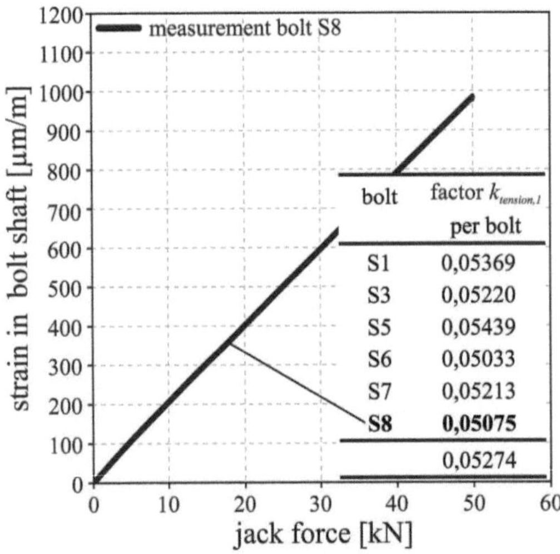

bolt	factor $k_{tension,1}$ per bolt
S1	0,05369
S3	0,05220
S5	0,05439
S6	0,05033
S7	0,05213
S8	**0,05075**
	0,05274

Figure 9:   Calibration of measurement bolts (here: measurement bolt S8), conversion of strain in measurement bolts into pre-tension force

In a second step a single rail support point was fixed to a concrete block using different grades of pre-stressing and surcharged by a cyclic compression-tension force ("Fig. 10").

Figure 10:   Test set-up with measurement devices (unstressed measurement bolt S4 and displacement transducers W2, W4)

The concrete block was designed in accordance to the actual *VFT-Rail*® girder (concrete strength, reinforcement, geometry). The measurement bolts were pre-stressed according to the tightening torque method whereas the final exact pre-stress was applied by means of the calibration factor $ktension,1$, monitoring the strain in the bolts. The set of measurement bolts used for this cyclic test was not re-used in the final bridge structure.

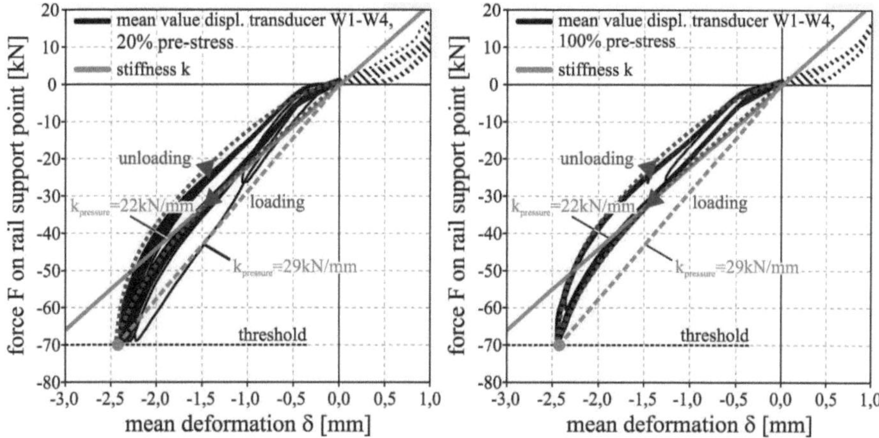

Figure 11: Mean lowering / deformation of the ribbed base plate depending on the force applied (20% pre-stress, 100% pre-stress)

The mean lowering of the ribbed base plate was measured by means of four displacement transducers and plotted against the jack force ("Fig. 11"). In this way the vertical stiffness of the rail support point structure under compression was determined. Within the elastic region of the elastic intermediate layer this stiffness equals the stiffness determined by TU Munich (research report N° 1689a, see chapter 3.1) as well as the stiffness used for design ($kpressure$ =22.5 kN/mm). However, due to the non-linear force-deformation behavior of the elastomer for compression forces > 45 kN, the secant stiffness ($kpressure$=29kN/mm) at $F =$ 70 kN was used within the scope of the monitoring, leading to slightly conservative values for small deformations.

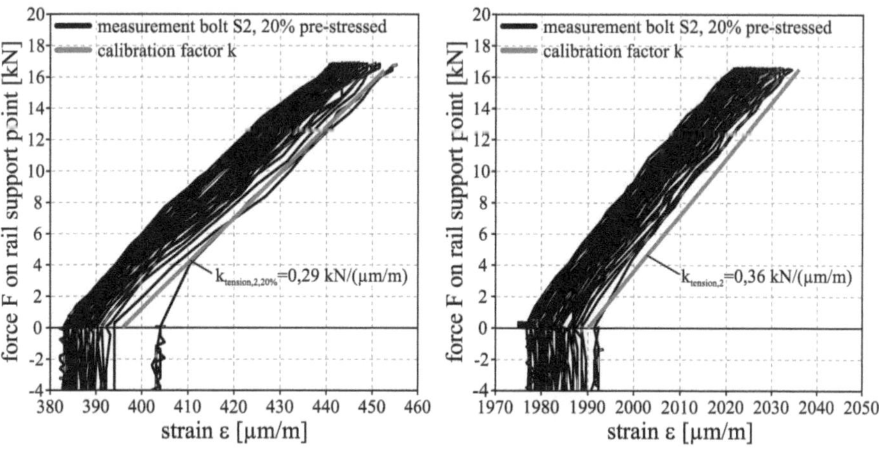

Figure 12: Strain change in measurement bolt S2 depending on applied force (20% pre-stressed, 100% pre-stressed)

The change in strain measured in bolts is plotted against the jack force for different degrees of pre-stress ("Fig. 12"). Due to design and function of the rail support point, compressive loading does not lead to a measurable decrease of strain in the measurement bolts. The strain-force relation $k_{tension,2}$ =0.36 [kN/(μm/m)] is valid for preloaded bolts only. Furthermore the linear relation is merely validated for tension forces Z<16 [kN]. A decrease of pre-stress results into a reduced factor $k_{tension,2,x\%,}$ whereas $kt_{ension,2}$ =0.36 [kN/(μm/m)] can be applied conservatively even for measurement bolts with a low pre-stress (20%). For non-preloaded measurement bolts, $k_{tension,1}$ (2 · 0,05274) may be applied ("Fig. 13" and "Fig. 9").

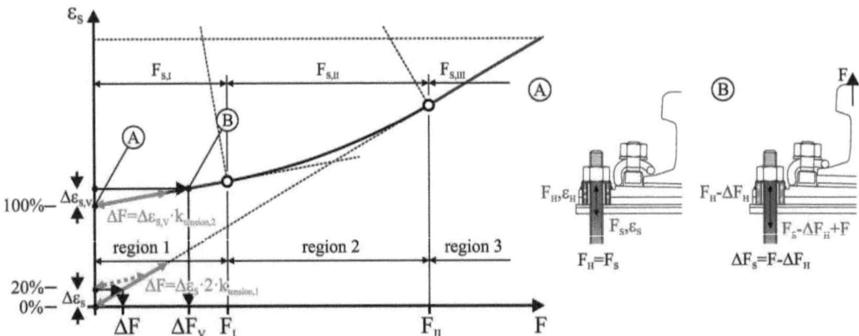

Figure 13:   Bolt force function: measured strain De$_s$ over jack force ΔF at rail support point

The following factors result for the determination of the forces acting on the rail support point:

■   Tension loading:

$$\Delta F_{tension}[kN] = \Delta\varepsilon[\mu m/m] \cdot k_{tension,2}\left[\frac{kN}{\mu m/m}\right] = \Delta\varepsilon \cdot 0.36 \qquad \text{(100\% pre-loaded bolts)}$$

$$\Delta F_{tension}[kN] = \Delta\varepsilon[\mu m/m] \cdot 2 \cdot k_{tension,1}\left[\frac{kN}{\mu m/m}\right] = \Delta\varepsilon \cdot 0.105 \quad \text{(non-preloaded bolts)}$$

■   Compressive loading:

$$F_{pressure}[kN] = \bar\delta[mm] \cdot k_{pressure}\left[\frac{kN}{mm}\right] = \bar\delta \cdot 22$$

The calibrated measurement bolts were installed during the construction of the railtracks. The pre-tension was applied according to the tightening torque method. The actual force in the bolts was determined based on the strain measured in the bolts.

3.2.2   Measurement results

Both the first and the fourth rail support point were monitored to verify the theoretically determined values of compression and tension forces acting on the system (section 3.1). On the one hand, strain changes in the measurement bolts were recorded to allow for a calculation of tensile forces acting on the rail support point. On the other hand the deformation of the ribbed base plate was measured to calculate the compressive force acting on the system. All short-term measurements were performed at a frequency of 1000 Hz.

The mean values of static forces recorded during a stop-run of the test train are plotted over the position of the train on the bridge for rail support point 1 and 4 ("Fig. 14"). These measured values correlate well with the characteristic calculated values (section 3.1). It can therefore be concluded that the design model chosen can be considered as sufficient.

A comparison of these results with the values recorded during "dynamic" train crossing at higher speeds shows that in the case of the present structure the velocity of the train crossing the bridge does not influence the forces acting on the rail support points. A dynamic amplification of forces could not be observed.

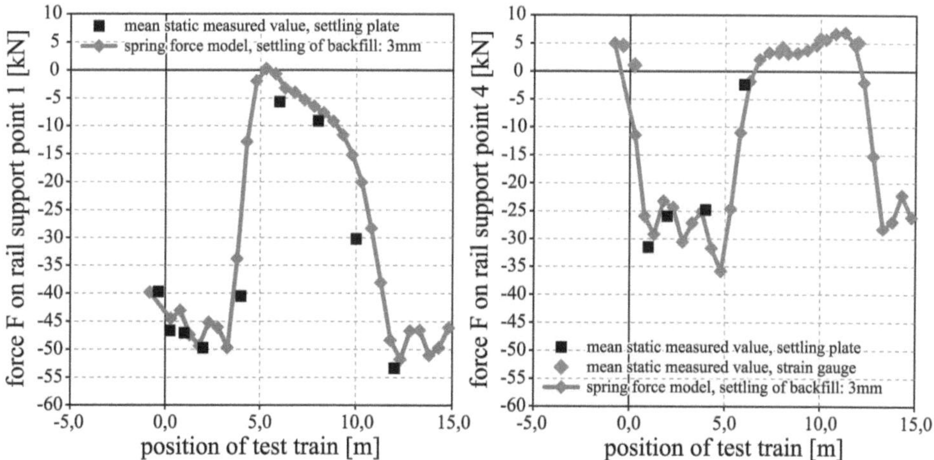

Figure 14: Compression force / tension force acting on controlling rail support points 1 (left) and 4 (right), caused by static loading (test train)

4. Local stresses in external reinforcement

4.1 Determination of design values

The internal forces needed for the determination of stresses were determined by means of a framework model. The section stiffness was calculated according to elasticity theory for condition I (uncracked concrete, "Fig. 15", left) as well as for condition II (cracked concrete in tension zone, "Fig. 15", right). In the case of condition II, tension stiffening was disregarded (cracked concrete not contributing to stiffness, Young's modulus of cracked concrete set to $E_c = 0$). In the region of compression, concrete action was considered with non-reduced Young's modulus. For the determination of the tension zone's height, neutral axis $z_{i,total} = 38.4$cm from condition I was considered without iteration. In both cases, next to external reinforcement, internal reinforcement (32 x Ø32) was taken into consideration as well. Especially in condition II, the influence of this so-called redundancy reinforcement on the total stiffness of the cross section could not be disregarded.

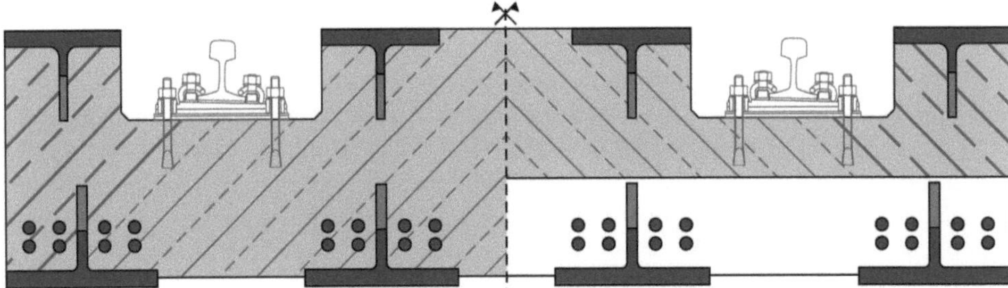

Figure 15:   Cross section for condition I (left) and condition II (right), rail support point drawn in for illustration purposes only

Internal bending moment M as well as internal shear force V were determined in the monitored sections for different load positions of test train BR 132 ("Fig. 5"). Therefore the loading of a single axle was applied step-wise to the framework model with a step-size of 1m, starting at axis 30 of the bridge. Based on these internal forces, normal stresses were calculated at the inner side of the upper and lower flanges, at the web as well as at the steel dowel (close to the stress hot-spot).

The stress calculation in the flanges as well as in the web was based on the global bending moment only. An amplification due to local stress concentration had not to be considered in this region, the stress amplification factor f_{global} was set to 1.0. The calculation of stresses in the region of the steel dowel was based on the global bending moment as well as on internal shear forces. There the stress amplification due to structural effects as well as local bending of the steel dowel (caused by internal shear forces) was considered by means of the stress amplification factors $k_{f,G,CL}$ and $k_{f,L;CL}$. Conservatively the stresses of both hot-spots were summed up [15] [16].

$$\Delta\sigma = k_{f,L} \cdot \frac{\Delta V \cdot S_y}{I_y \cdot t_w} + k_{f,G} \cdot \left(\frac{\Delta N}{A} + \frac{\Delta M}{I_y} \cdot z_D \right)$$

4.2   Field measurements

4.2.1   Test set-up and measurement

Due to symmetry of the *VFT-Rail®* girder, the strain gauges were applied at the four left steel girders (2 top-girders, 2 bottom-girders, "Fig. 17") between axis 20 and axis 30 of the bridge structure only.

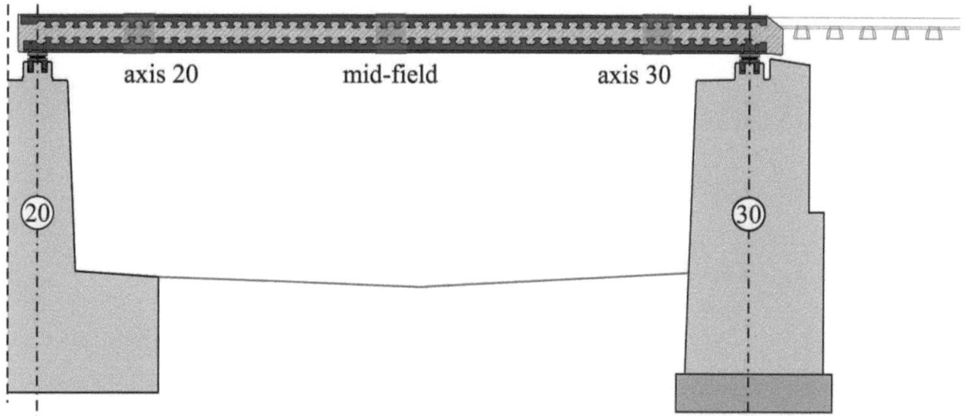

Figure 16: Measurement sections at superstructure between axis 20 and axis 30

In axial direction, three measurement sections were defined. On the one hand the position of the two sections close to the bearings (axis 20 and axis 30) had to be chosen in a way to allow for measurement of high shear forces in the composite section. On the other hand, local effects due to the load transfer into the bearings had to be minimized. Therefore the two outer sections were defined in a distance of about 1.85m to the bearing axis around the 7th and 8th steel dowel. The third measurement section was defined in mid-field around steel dowel 26 and steel dowel 27, as in this section the highest bending moments and deformations occurred ("Fig. 16").

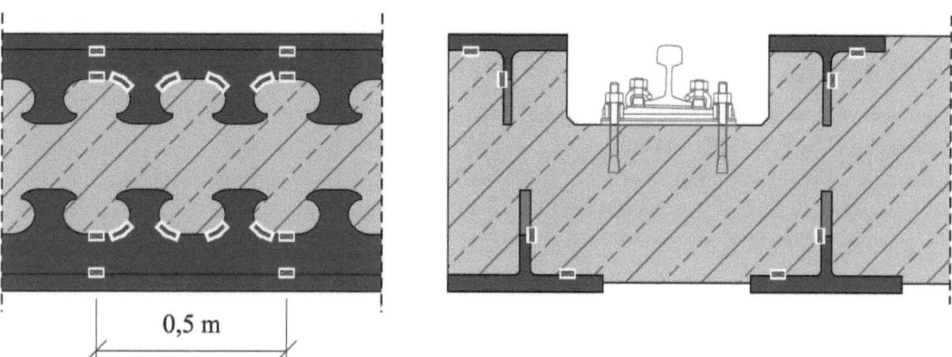

Figure 17: Position of strain gauges

The strain gauges were applied prior to casting in the workshop. In total 72 conventional steel strain gauges (R = 100 Ohm) were applicated to the steel girder (Tab. 1). As all strain gauges were covered by concrete, special care needed to be taken prior to casting regarding the protection of the gauges. Therefore all strain gauges were covered by a 2-component adhesive (PS-adhesive) after application of the signal wires. After hardening, this adhesive formed a stiff protective cover. Finally a layer of permanent elastic cement (AK22) was applied, which minimized the risk of the gauges to be pulled off due to concrete pressure ("Fig. 18"). Furthermore all signal wires were placed inside elastic ducts, placed in the radius of the steel profile and leading to axis 30 of the bridge structure. Due to these measures, only 4 out of 72 strain gauges did not work properly after casting, transportation and installation of the elements.

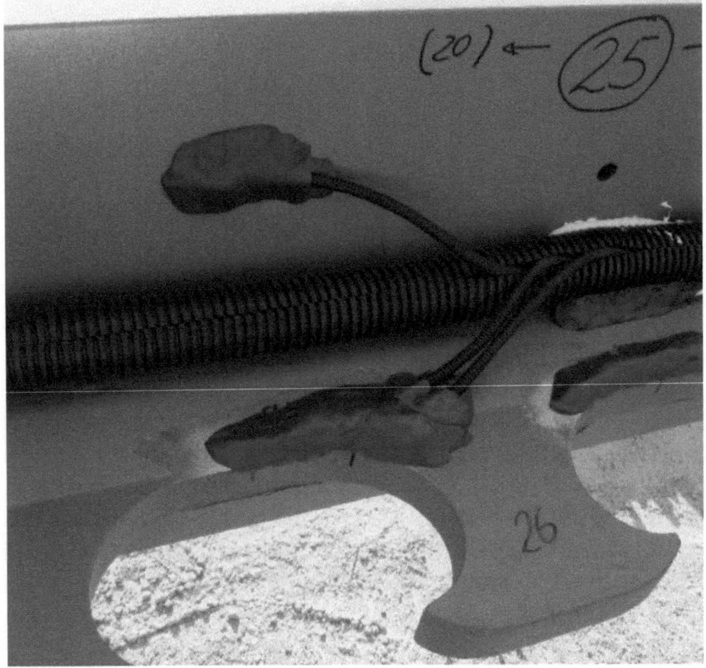

Figure 18:   Protected strain gauges, covered with 2-component adhesive (PS-adhesive) and permanent elastic cement (AK22)

In the three measurement sections ("Fig. 16") strain gauges were applied at the flanges and the web ("Fig. 17"). At the two outer sections close to the bearings, the strains were measured at a single upper and a single lower flange only to minimize the number of strain gauges. At the section in mid-field the gauges were applied at all four steel girders. The strain gauges placed at the flanges and at the web allowed for the determination of the linear strain distribution over the cross section. Furthermore the application of strain gauges in front of and behind the section under investigation ("Fig. 16") gave the possibility to calculate the shear forces introduced by a single steel dowel into the concrete.

Additional strain gauges were applied close to the hot-spots ("Fig. 17", left). Based on experiences gained in former research projects [17] [15] [18] the position of these strain gauges was chosen such to measure the maximum strains occurring in that region. There it has to be distinguished between two possible maximum strain distributions. On the one hand, a strain maximum occurs in the upper region of the clothoide due to shear forces acting at the steel dowel ("Fig. 19", left). On the other hand, a strain maximum occurs close to the base of the dowel, caused by axial deformation of the external reinforcement due to global bending ("Fig. 19", right). However, a simple prediction of the actual position of the relevant hot-spot is not possible, as both effects interfere, based on the position of the steel dowel as well as the position of the loading. Furthermore it is not possible to measure the strain at the actual hot-spot directly, as it always occurs directly at the cutting line. Therefore the strain gauges were applied at an angle of 24° and 41° in a distance of 5mm to the cutting line which covers the predicted hot-spot region. Finally the maximum strains / stresses were determined by extrapolation of the measured strains, amplified by the stress concentration factors $k_{f,L}$ and $k_{f,G}$ [3].

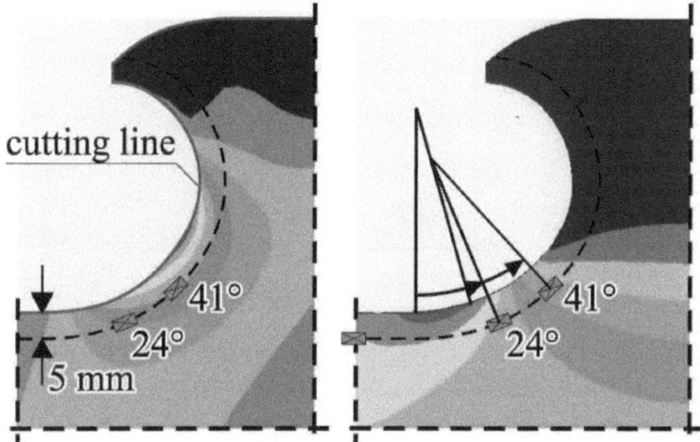

Figure 19: Calculated stress distribution (schematic, left: local loading due to shear force acting at steel dowel, right: global loading due to axial deformation of the external reinforcement due to bending of the composite section) [15].

4.2.2 Measurement results

Within the scope of this paper, selected results of the strain measurements are discussed. Therefore the strains measured are transformed into stresses. At first, results are presented gained by measurement of strains at flanges, web and within the hotspot region (see chapter 4.2.1) during the stop-runs of the test train BR 132 ("Fig. 5"). These results are compared with the theoretic values (section 4.1). Afterwards the strain values recorded during crossing of the test train are compared to the strain values measured for the scheduled passenger trains ("RegionalExpress") DBAG 612 ("Fig. 6").

In Fig. 20 the strains on the inner side of the flange (see "Fig. 17" for exact location) are compared to the theoretic characteristic values. The stresses are plotted over the bridge length; the single measurement sections (axis 30, midfield, axis 20) can be clearly identified. The differently colored single points represent the stress values for different load positions of the test train (position of first train axle relative to axis 30 of bridge structure). The colored curves represent the envelope for the respective load position, whereas the red curves represent the limiting envelope of all load positions.

The parabolic distribution of results over the length of the single girder indicates that mainly normal stresses caused by global bending action are present in the tension flange. A comparison of the measured values to the theoretic values of condition I shows, that the measured strains / stresses exceed the characteristic design envelopes mainly in mid-field ("Fig. 20", left). This leads to the conclusion that cracking of the concrete occurred in the section's tension zone, redistributing the stresses to the structural steel. However, all measured values are situated clearly below the characteristic design envelopes for condition II ("Fig. 20", right).

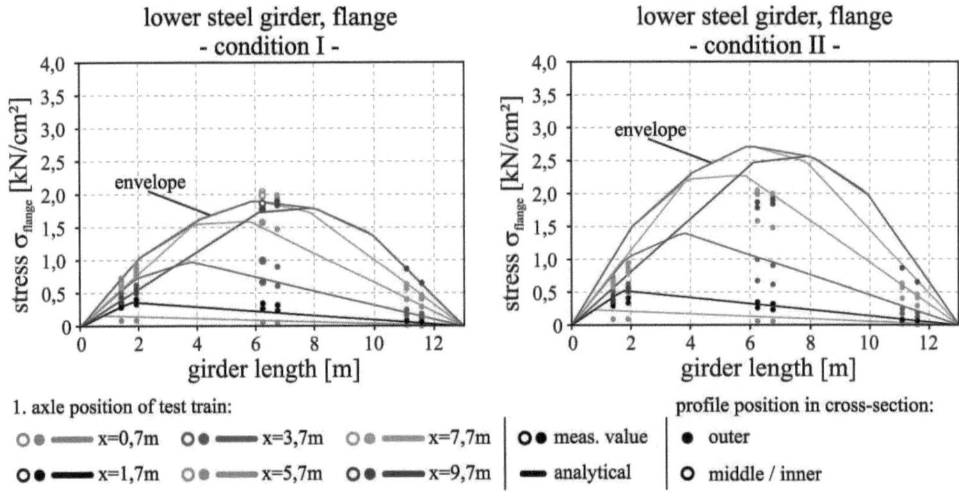

Figure 20:   Measured strains / stresses on the inner side of the flange (lower external steel girder, points) compared to calculated stresses for condition I (left) and condition II (right)

The strains measured in the web are also distributed parabolically over the length of the girder ("Fig. 21"). This again leads to the conclusion that mainly normal stresses caused by global bending action are present in the tension flange. These stresses in the web are by 25% lower than the corresponding stresses in the flange, which can be attributed (amongst others) to the smaller lever arm to the neutral axis. All values are increased compared to the theoretic values of condition I, which again shows that condition II governs and needs to be considered during design.

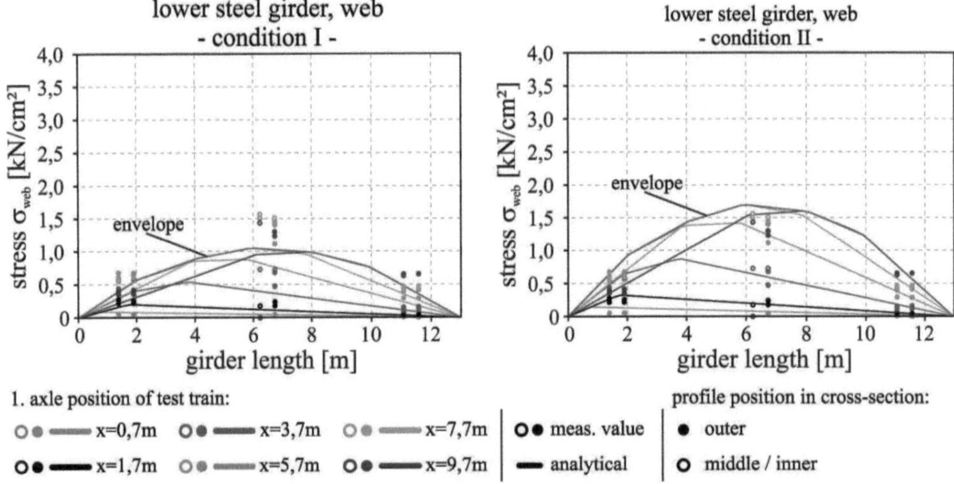

Figure 21:   Measured strains / stresses at web of lower external reinforcement, compared to calculated values for condition I (left) and condition II (right)

Stresses due to global bending of the composite girder and stresses caused by local bending of the steel dowel superimpose within the region of the socalled hot-spot at the foot of the dowel ("Fig. 22", "Fig. 19"). Furthermore the stresses are increased due to geometric notch effects. Close to the bridge's bearings (axis 20 and axis 30) normal stresses due to local bending of the steels dowels, caused by shear forces to be distributed by the dowels, predominate. In mid-field, global bending still governs the stress state. However, as the stresssums of global and local stress have to be taken into account, the hotspots both at midfield as well as at the bearings have to be investigated to find the critical crosssection. In the case given here, the critical stresses can be found close to the bearings.

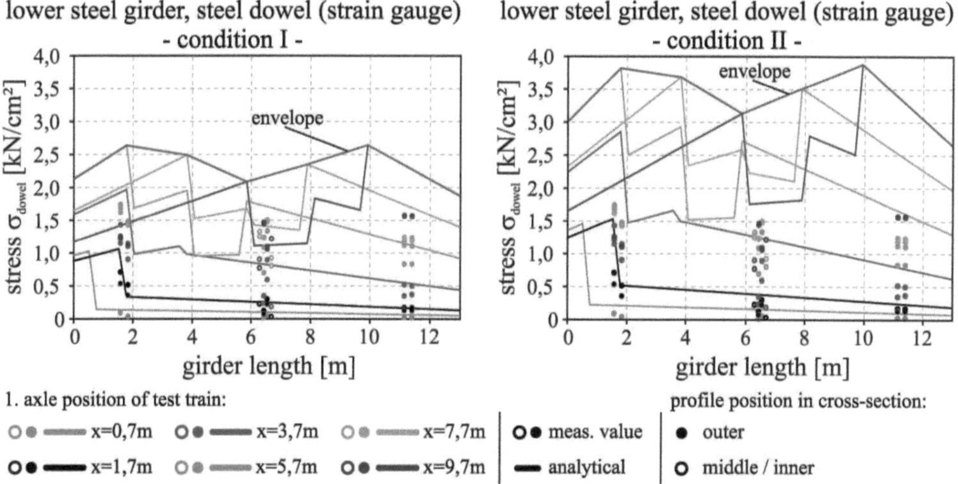

Figure 22:   Measured strains / stresses within the region of the hot-spot, lower external reinforcement, compared to calculated values for condition I (left) and condition II (right)

The results presented here show, that the maximum tension stress of about 2 kN/cm² can be measured in mid-field in the region of the lower flanges. Therefore the stresses in this region (strain gauge S45 and strain gauge S46) induced by the test train (100 km/h) are compared to those caused by the scheduled regional train RE 3343 (120 km/h). The crossings lasted about 1 s (BR 132) and about 2 s (RE 3343) respectively ("Fig. 23"). The bogies are visible in both plots, whereas the two central bogies of BR 132 merge due to the short distance between them (comp. "Fig. 6"). Furthermore the stresses caused by the regional train prove to be clearly smaller (about 0,87 kN/cm²) than the ones caused by the test train ("Fig. 23", left).

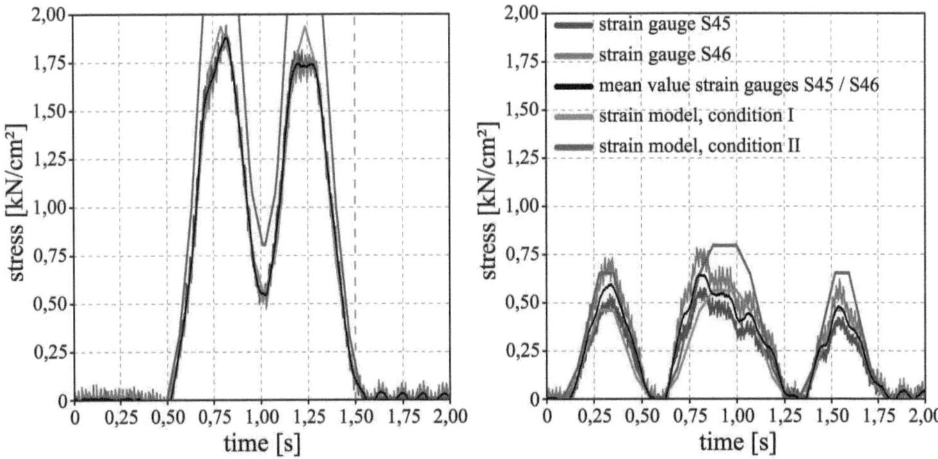

Figure 23:   Stresses in lower flange caused by crossing of test train (100 km/h, left) and regional train (120 km/h, right)

Furthermore the crossing of the regional train (BR 132) with 120 km/h is compared to a calculated quasi-static crossing. Therefore the estimated total weight of 58 t is distributed evenly to all axes, which results in axle loads of 72.5 kN. Based on the cross sectional values, the resulting moment distribution is transformed into theoretic stress curves for condition I (orange line) and condition II (red line) ("Fig. 23"). These theoretic values show a good correlation to the measured ones. The gliding mean value of both measured values (strain gauge S45 and strain gauge S46) is situated between the curves of condition I and condition II, which corresponds to the results discussed before. The small deviations in time between measured values and hand calculation results can be attributed to dynamic effects.

It can be concluded, that there is a good correlation between calculated characteristic stresses and measured stresses (section 4.1). Together with the evaluation of results in the region of the hot-spot, this leads to the conclusion that the given design concept and safe-sided design rules based on stress concentration factors is applied correctly.

5.   Outlook

The successful realization of the *VFT Rail*® bridge in Simmerbach, combined with the prove of the applicability of state of the art calculation methods by monitoring measures lead to an extension of approval of this new bridge type in Germany [19]. In addition the general technical approval for composite dowels [3] was adopted in 2013, which allows for an economic, timesaving and simpler application of these shear connectors in future.

References

[1]   Seidl, G. et al., "PRECOBEAM: Prefabricated enduring composite beams based on innovative shear transmission", Final Report RFSR-CT-2006-00030, Research Fund for Coal and Steel, Brussels, 2013.

[2]   Seidl, G., Mensinger, M., Koch, E., Hugle, F., „Projektbericht Eisenbahnüberführung Simmerbach – Pilotbrücke in VFT-Rail Bauweise mit externer Bewehrung," Stahlbau 81, pp. 100-107, 2012.

[3]   Allgemeine Bauaufsichtliche Zulassung, Verbunddübelleisten, Z-26.4-56, Deutsches Institut für Bautechnik DIBt, Berlin, 2013.

[4]   ThyssenKrupp GfT Gleistechnik, Oberbauhandbuch, Bochum, 2006.

[5]   „ECOBRIDGE: Demonstration of economical bridge solutions based on innovative composite dowels and integrated abutements," Final Report RFSP-CT-2010-00024, Research Fund for Coal and Steel, Brussels, not published yet

[6]   Rauert, T., Bigelow, H., Hoffmeister, B., Feldmann, M., Patz, R., Lippert, P., „Zum Einfluss baulicher Randbedingungen auf das dynamische Verhalten von WIB-Eisenbahnbrücken, Teil I," Bautechnik 87, pp. 665–672, 2010

[7]   Rauert, T., Bigelow, H., Hoffmeister, B., Feldmann, M., Patz, R., Lippert, P, „Zum Einfluss baulicher Randbedingungen auf das dynamische Verhalten von WIB-Eisenbahnbrücken, Teil 2," Bautechnik 87, pp. 751–760, 2010.

[8]   Rauert, T., Bigelow, H., Hoffmeister, B., Feldmann, M, „On the prediction of the interaction effect caused by continuous ballast on filler beam railway bridges by experimentally supported numerical studies," Engineering Structures, vol. 32, pp. 3981-3988, 2010.

[9]   DB Netz AG, „Bemessungsgrundlagen für die Festhaltekonstruktionen der Festen Fahrbahn auf Brücken," Ril804.5404, 2011.

[10] DB Netz AG, „Nachweise an den Überbauenden bei Fester Fahrbahn," Ril 804 Modul 5405 (draft), 2011.

[11] DIN FB 101, Beuth Verlag GmbH, Berlin, 2009.

[12] Eisenmann, J., „Gleislagebeständigkeit des Schotteroberbaus," Der Eisenbahningenieur, vol. 2, pp. 20-26, 2011.

[13] Feldmann, M., Naumes, J., Pak, D., „Zum Last-Verformungsverhalten von Schrauben in vorgespannten Ringflanschverbindungen mit überbrückten Klaffungen im Hinblick auf die Ermüdungsvorhersage," Stahlbau 80, pp. 21-29, 2011.

[14] Veljkovic, M., Feldmann, M., Naumes, J., Pak, D., Rebelo, C., Simões da Silva, L., „Friction connection in tubular towers for a wind turbine," Stahlbau 79, pp. 660-668, 2010.

[15] Feldmann, M., Hegger, J., Gündel, M., Kopp, M., Gallwoszus, J., Heinemeyer, S., Seidl, G., Hoyer, O., „P804 – Neue Systeme für Stahlverbundbrücken – Verbundfertigteilträger aus hochfesten Werkstoffen und innovativen Verbundmitteln. VERBUND-DÜBEL-LEISTEN," FOSTA – Forschungsvereinigung Stahlanwendung e. V., Düsseldorf, not published yet.

[16] Mensinger, M., „Gutachterliche Stellungnahme G 1008 Low Cost Bridge: Beurteilung der Konzepte zum Nachweis der statischen Tragfähigkeit und der Ermüdungssicherheit der Verdübelung der externen Bewehrung des neuartigen Brückentyps „Low Cost Bridge"," Munich, 2010.

[17] Feldmann, M., Hegger, J., Hechler, O., Rauscher, S., „P621 – Untersuchungen zum Trag- und Verformungsverhalten von Verbundmitteln unter ruhender und nichtruhender Belastung bei Verwendung hochfester Werkstoffe," FOSTA – Forschungsvereinigung Stahlanwendung e. V., Düsseldorf, 2007.

[18] Heinemeyer, S., „Zum Trag- und Verformungsverhalten von Verbundträgern mit Puzzleleisten und ultrahochfestem Beton", dissertation, Aachen, 2011.

[19] Eisenbahnbundesamt, „Erweiterung der Zulassung zur Betriebserprobung vom 06.12.2010 der Bauart VFT-Rail der SSF Ingenieure AG zur Verwendung in Strecken der Eisenbahn des Bundes mit zulässigen Geswchwindigkeiten bis 160 km/h", Bonn, 2013.

Modern composite bridges by VFT-WIB method in Poland realized at new express road S7 at Olsztynek-Nidzica sector

Wojciech Lorenc[1] & Tomasz Kołakowski[2]

Keywords: VFT-WIB, composite bridges, crossings for animals, composite dowels.

Abstract: The design-an-build formula allowed for several new bridges, including road bridges and animal crossings, to be built using the VFT-WIB solution for the road S7 at Olsztynek – Nidzica in Poland. The designers took advantage of the versatility and adaptation capability of VFT-WIB, overrunning some of the encountered situations, which include the difficult geometry or the contractor's demands.

1.   Introduction

The new express road S7 at Olsztynek-Nidzica sector in northern part of Poland was realized by design-and-build formula and the contract was conducted between the years 2009 and 2012. The works included the design and build of two new lanes of the road S7 at Olsztynek – Nidzica, sector from km 175+800 to km 203+600 and bypass of Olsztynek (national road no. 51) from km 109+500 to km 115+500.

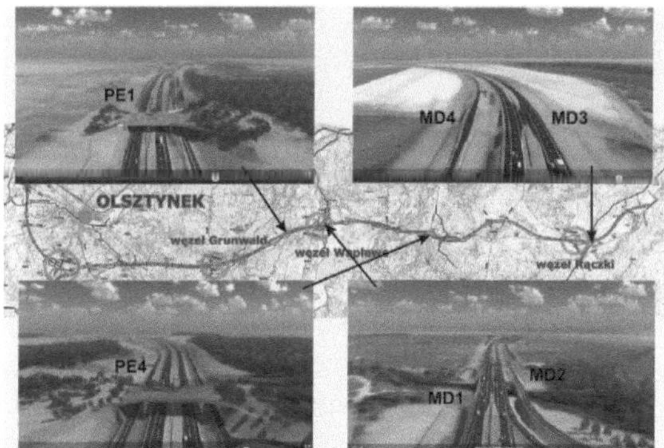

Figure 1:   Naming and visualization of structures realized in course and over S7 road at Olsztynek – Nidzica sector

1   Dr. Ing., Wrocław University of Technology, Wybrzeze Wyspianskiege 27, Wrocław, Poland, wojciech.lorenc@pwr.wroc.pl
2   Dipl. Ing., EUROPROJEKT GDANSK S.A., Nadwislanska 55, Gdansk, Poland, europrojekt@europrojekt.pl

Several bridges and crossings for animals were realized with the VFT-WIB method, next to the typical VFT solution used for 12 integral bridges. The experiences gained from this contract allowed for the VFT-WIB method to be treated as a typical solution for bridges in Poland nowadays. There were 6 different structures realized in this technology [1] including road bridges (MD1, MD2, MD3, MD4) and crossings for animals (PE1, PE4) which are presented in "Fig. 1". Shear connection by composite dowels is used [2, 3] for all structures being presented with its specific MCL250/115 shape [4].

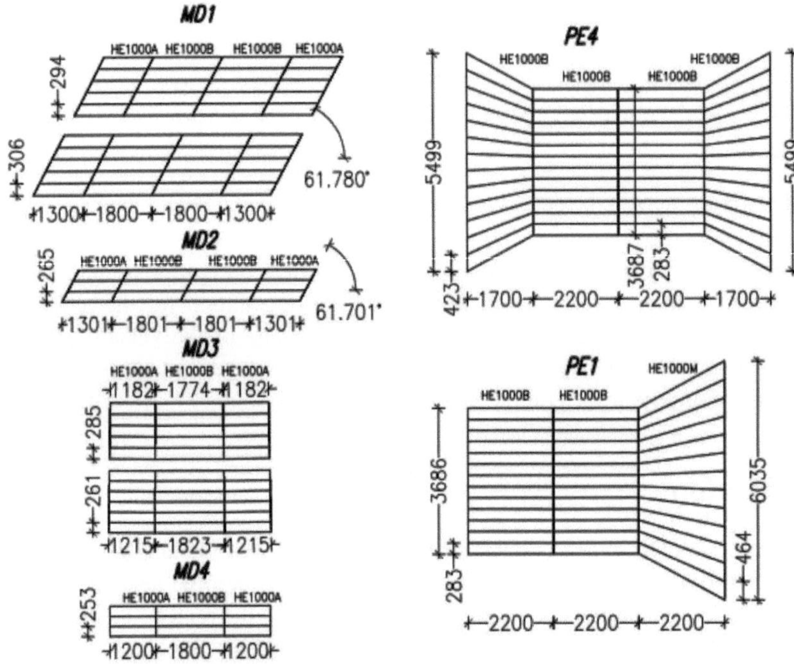

Figure 2: Girder systems of the structures realized in course and over S7 road at Olsztynek – Nidzica sector

2. Road bridges

The road bridges (MD) consist of independent structures for single lanes and are designed as continuous beams. The crossings for animals are realized as integral structures. The total tonnage of steel built in the structures exceeded 1100 tons of S355M steel in HE1000A, HE1000B, HE1000M sections. Road bridges in course of S7 can be treated as system solution – they are designed using double VFT-WIB section and constant height of section, of 583 mm for the prefabricated elements and of 250 mm for the in-situ slab, with a total height of the composite beam of 833 mm. The steel shapes HE1000A and HE1000B out of S355M steel were used, depending on the span length which was up to 18 m. The width of the prefabricated elements and the number of beams in the superstructures varied, according to the geometry of the individual structures. The maximal width of the prefabricated elements was 3060 mm. MD1 bridge consists of two structures of 18.44 m and 17.70 m width, the spans are L=13+18+18+13m, the skew angle is 62 degrees. There are 6 beams in the transversal section of each structure. MD2 bridge consists of a single structure of 10.70 m width, the spans are L=13+18+18+13m, the skew angle is 62 degrees. There are 4 beams in the transversal section of the structure. MD3 bridge consists of two structures of 17.19 m and 18.35 m width, it is curved in plane and the spans are approximately L=12+18+12m – lengths of individual beams varied due to the arching in plane. MD4 bridge consists of a single structure of 10.20 m width, the spans are L=12+18+12, there is no skew angle. There are 4 beams in the transversal section of the structure.

The transversal section of a single structure (under single lane) of MD1 bridge is presented in "Fig. 3" (the other bridges are analogical). The prefabricated girders are shown in "Fig. 4" and "Fig. 5". In "Fig. 6" the final structures MD1 and MD2 are presented.

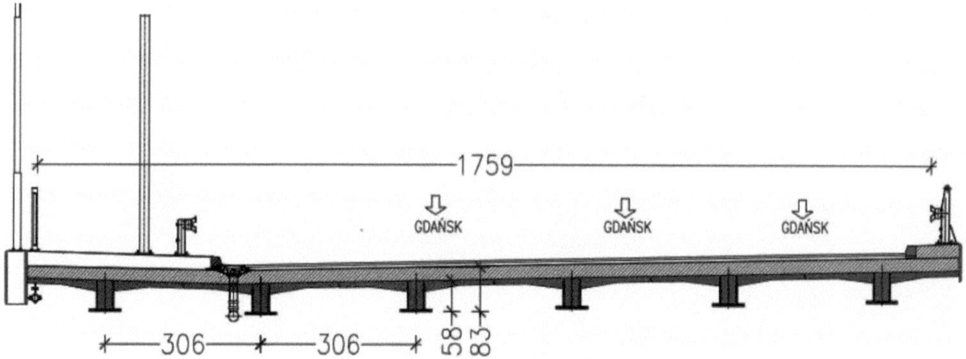

Figure 3:   Transversal section of single structure of MD1 bridge

Figure 4:   Prefabricated girder of road bridge on site

Figure 5: Prefabricated girder on temporary supports

Figure 6: Bridges MD1 and MD2 [picture from: http://s7.olsztynek-nidzica.zaprojektuj-zbuduj.pl/roboty-mostowe.html]

3.   Crossing for animals PE4

Much more complex solutions, in comparison to the road bridge system, are applied for animal crossings over the S7 road. There are two such structures realized, signed PE1 and PE4. In comparison to road bridges the load for animal crossings is much heavier but on the other hand no fatigue problems appear. The PE4 structure is presented at first, its realization was supported by international Ecobridge RFCS project (Demonstration of ECOnomical bridge solutions based on innovative composite dowels and integrated abutments, contract RFSP-CT-2010-00024 of Research Fund for Coal and Steel). However, the geometry of the structure is complicated, halved HEB1000 sections have been used for the steel part of the entire composite structure. They have been locally strengthened by coverplates at supports.

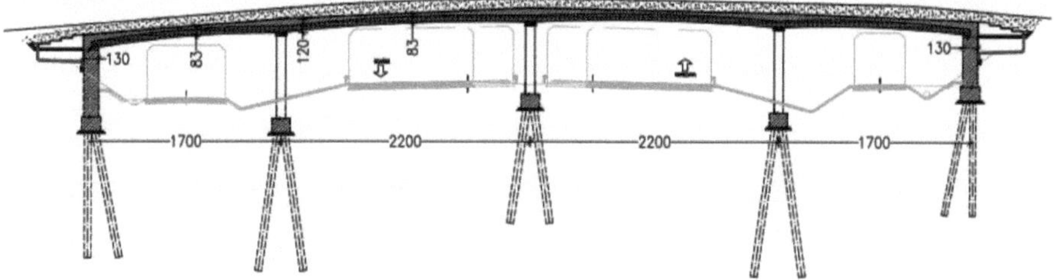

Figure 7: Longitudinal section of PE4 structure

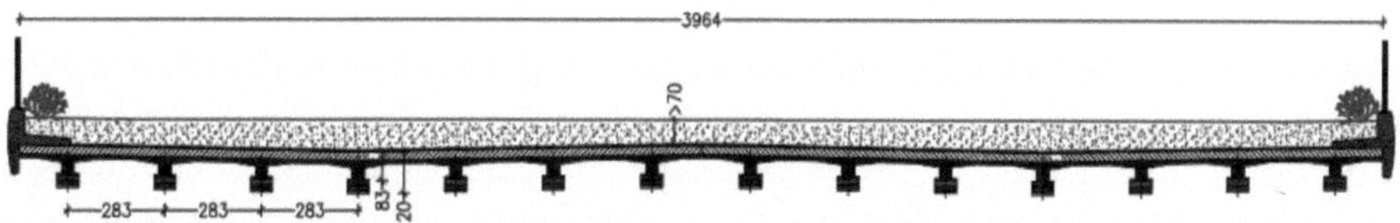

Figure 8: Transversal section from a 22m span of PE4 structure

The structure ("Fig. 7", "Fig. 8") is a 4-span frame with variable height of girders and, moreover, with variable width of external spans (fan-shaped spans). The entire structure consists of two identical substructures (mirroring) divided by longitudinal dilation between them. The geometry of the individual structures is presented in "Fig. 9" by using visualization of FE model used for design. The idea for efficient design was to obtain variable height of girder using constant height of prefabricated element ("Fig. 10").

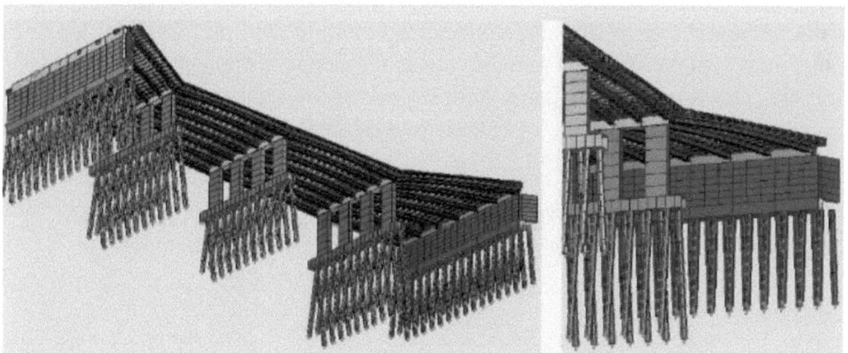

Figure 9: Geometry of PE4 structure – visualization of FE model of single substructure

Figure 10: Idea for variable height of girder used in PE4 structure

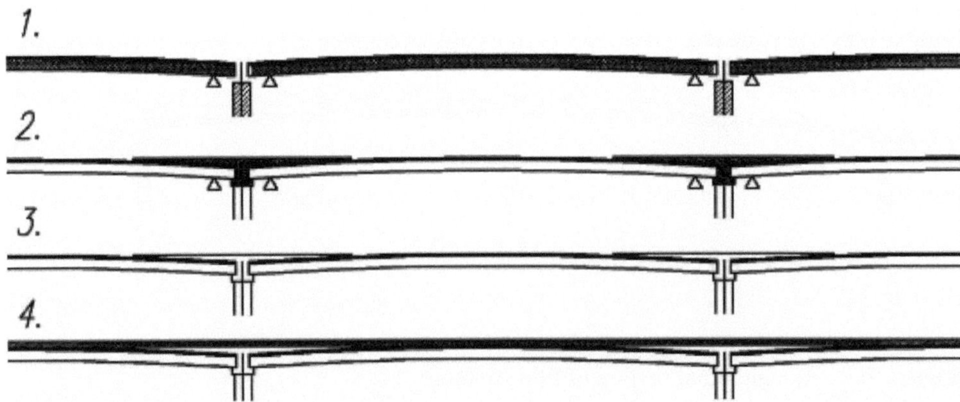

Figure 11:   Technology of realization of superstructure of PE4

It leads to the situation where the level of the neutral axis of the composite section is next to the shear connection level. To form such a structure the casting of crossbeams is made together with an additional layer of concrete between constant in-situ slab and prefabricated slab. The additional concrete layer appears only at the support region and it has variable height increasing towards support. This way bigger lever arm can be reached for handling the load of the in situ slab, reinforcement in this layer does not have to be checked for crack width and no additional step in technology is needed (as this layer is casted as a part of the crossbeam). Moreover, increased concrete slab thickness in support region leads to better cooperation between girders due to increased torsional stiffness of slab ("Fig. 11").

According to principia of the VFT-WIB method, prefabricated composite elements were initially assumed to be used for construction of superstructures. The complicated geometry of the bridge led to a very complicated geometry of the prefabricated elements. Especially, prefabricated girders for spans 1 and 4 differ much, depending on their location in the span. After many discussions the general contractor decided to build the first line of bridge in its final position, contrary to the assumed prefabrication of composite elements at ground level next to the bridge. This way steel elements were supported the same way as it was initially designed (5 points along the girder) but using high towers typical for in-situ implementations. It was possible because there was no traffic under the bridge.

Figure 12:   Fabrication of steel structure: a) separation of HEB1000 sections, b) single T-shape and tandem shape, c) additional coverplate welded at support region

Figure 13: Foundations

Figure 14: Reinforcement cages of pillars

Figure 15:   Steel shapes at front and girders of first line of the bridge in final position in the background

This way the realization of prefabricated elements took place in final position ("Fig. 15"). The reinforcing works of the girders could be made in parallel to the fabrication of crossbeams. The welded connections of the steel girders using additional steel plates were assumed instead of screwed connection usually used for VFT confirmed to be a good solution, as the problem of tolerances (studied in detail at the design stage) really appeared. It was easy to handle by welding, but it could be a real problem by screws. Moreover, it was a big problem to realize the reinforcement of crossbeams (which was studied at the design stage also) – due to aesthetics relatively small crossbeams and pillars were used ("Fig. 16", "Fig. 17"). This way many bars in the transition zone between pillar and crossbeam was a hard task to handle.

Figure 16:   Reinforcement of the top of pillar to be casted in the crossbeam

After the prefabricated plates of the girders had been casted ("Fig. 18"), intermediate temporary supports have been removed, to get the same layout of strains as for prefabricated girders shifted by crane and put in their final position on temporary supports next to the pillar. Then, reinforcement of the additional concrete layer (about 5 meters from the support axis) to be casted together with crossbeams ("Fig 17") has been implemented and the elements were casted.

Figure 17: Reinforcement of the main girders and crossbeam (fit in reinforcement of pillars) prepared in their final position

Figure 18: First concrete slab of the main girders prepared in their final position

Finally, reinforcement for the last in-situ slab was placed and the concrete was casted. The second line of the bridge was not realized after the first line was finished, but almost in parallel. During the execution of the first line of the bridge first experiences have been gained and discussions appeared concerning the execution technology of the second line of the superstructure. Finally it was decided, that the initially designed execution technology (prefabricated girders realized next to the bridge and then shifted by crane for final positions in spans) is possible even for such a complicated geometry of the girders. Prefabrication stands were set next to the bridge ("Fig. 19") in few locations because the number of girders to be produced was large. Girders for spans 2 and 3 were executed possibly close to their final locations and girders for spans 1 and 4 were realized behind the abutments. Prefabrication is presented in the part of manuscript concerning PE1 structure and all girders for PE1 have been produced this way.

Figure 19:   Prefabrication of girders for second line of PE4 structure

The reinforcement system of the prefabricated elements had to be orthogonal for side spans, as it is presented in "Fig. 20" and "Fig. 21". This way each beam of side span was individual one.

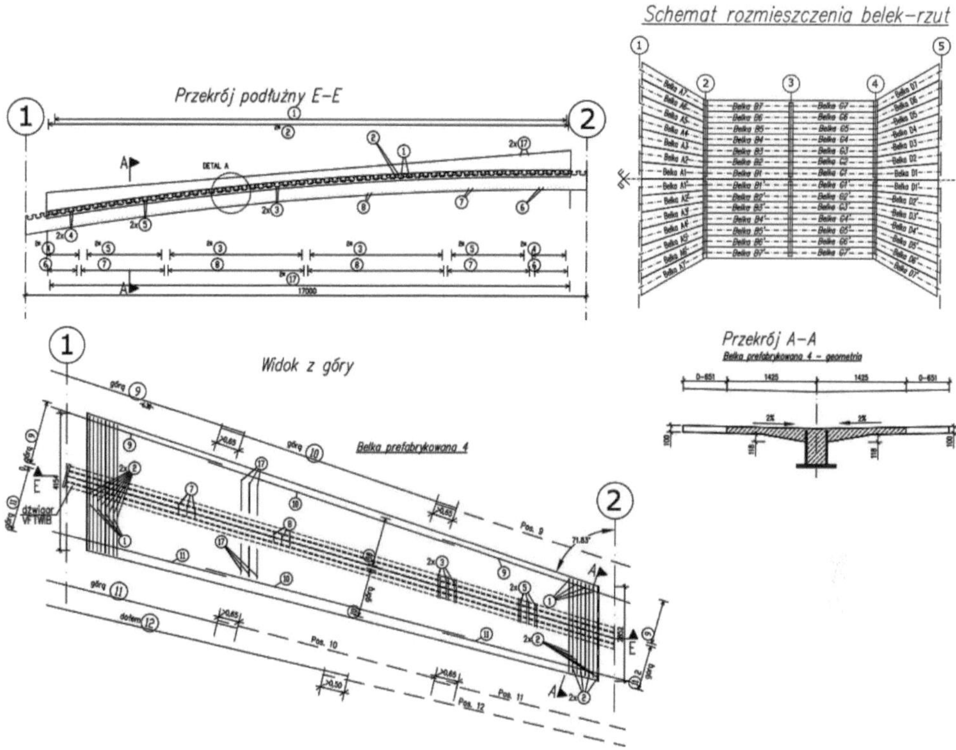

Figure 20: Geometry and reinforcement system of A4 beam for side span of PE4 structure

Figure 21: Realized reinforcement system of A4 beam for side span of PE4 structure

The prefabricated girders were shifted by crane to their final positions ("Fig. 22") on temporary supports next to pillars and abutments. Temporary supports were used to realise scaffolding for crossbeam and to support girders for intermediate construction stage.

Figure 22: The prefabricated girders of the second line of PE4 structure in their final positions

Figure 23: PE4 structure before construction of embankments

Figure 24:   PE4 structure before construction of embankments

Finally prefabricated elements of cornice ("Fig. 25") have been realized next to structure and placed on superstructure and soil layers have been applied. Realized bridge is presented in "Fig. 26".

Figure 25:   Prefabricated elements of cornice

Figure 26:   General view of crossing for animals PE4

4.   Crossing for animals PE1

PE1 ("Fig. 27") bridge is a frame with three spans (22m+22m+22m) and it differs comparing to the PE4 structure: there is only one fan-shaped span, but it is bigger comparing to PE4. This resulted in very big width of the prefabricated elements of this fan-shaped span. The bridge is founded on bored piles ("Fig. 28") because of weak soil conditions, quite different comparing to PE4.

Figure 27:   General view of crossing for animals PE1

Figure 28:   Visualization of computational model of the bridge

This structure was realized in parallel to the second part of PE4 and full prefabrication was used ("Fig. 29" to "Fig. 32"). The large width of the prefabricated elements is evident ("Fig. 33") and variable along the girder for the fan-shaped span. The large area of the bridge resulted in large areas used for prefabrication of girders next to the bridge ("Fig. 34").

Figure 29:   Steel shapes at prefabrication stand

Figure 30:   Scaffolding of prefabricated VFT-WIB girders

Figure 31:   Reinforcement (at right) and realized (at left) prefabricated VFT-WIB girders

Figure 32:   General view of prefabricated girders for fan-shaped span of PE1 (right) and straight girders for simple spans (left)

Figure 33:   Large width of prefabricated girders for fan-shaped span

Figure 34:   General view of construction site when full prefabrication used

Figure 35:   Bottom view of girders at finished P11 structure

5.   Acknowledgement

The design and construction of PE4 was carried out in frame of the international project "Demonstration of ECOnomical bridge solutions based on innovative composite dowels and integrated abutments" financed by the European Commision, contract RFSP-CT-2010-00024 of Research Fund for Coal and Steel and with financial support of the MNiSW of the Polish government in frame of the contract 2219\FBWiS\2011\2.

References

[1]   Seidl G., Stambuk M., Lorenc W., Kołakowski T., Petzek E., "Wirtschaftliche Verbundbauweisen im Brückenbau – Bauweisen mit Verbunddübelleisten", Stahlbau 82, Heft 7, 2013

[2]   Berthellemy J., Lorenc W., Mensinger M., Rauscher S., Seidl G., "Zum Tragverhalten von Verbunddübeln, Teil 1: Tragverhalten unter statischer Belastung", Stahlbau 80, Heft 3: 172-184, 2011

[3]   Berthellemy J., Lorenc W., Mensinger M., Ndogmo J., Seidl G., "Zum Tragverhalten von Verbunddübeln, Teil 2: Ermüdungsverhalten", Stahlbau 80, Heft 4: 256-267, 2011

[4]   Lorenc W., Kożuch M., Seidl G., "Zur Grenztragfähigkeit von Verbunddübeln mit Klothoidenform", Stahlbau 82, Heft 3: 196-207, 2013.

Bridges by VFT method in Poland: state-of-the-art

Tomasz Kołakowski[1] & Wojciech Lorenc[2]

Keywords: VFT, composite bridges, integral bridges, semi-integral bridges.

Abstract: Bridges with VFT method have been built in Poland since the year 2000. Many diverse solutions and structures were developed, each adapting to the different conditions and requirements. The experience gained in time and the well behavior of the VFT bridges, allowed for the solution to be appreciated.

1.   Introduction

After the year 2000 bridges using the VFT (VerbundFertigteilTräger – prefabricated composite beams) method have been introduced in Poland by Europrojekt Gdańsk design office with support of SSF. The first bridge was constructed in Poland in the year 2004 after the successful implementation of VFT construction method in Germany. Starting from the static scheme of continuous beams and viaducts constructed mainly above railway lines ("Fig. 1") the construction method by prefabricated composite girder evolved to semi integral and integral structures.

Figure 1:   Two-span continuous beam (43m+43m) VFT viaduct next to Kije

1   Dipl. Ing., EUROPROJEKT GDANSK S.A., Nadwislanska 55, Gdansk, Poland, europrojekt@europrojekt.pl
2   Dr. Ing., Wrocław University of Technology, Wybrzeze Wyspianskiege 27, Wrocław, Poland, wojciech. lorenc@pwr.wroc.pl

2.   Solutions by VFT method used in Poland

The typical cross-section of the superstructure of a VFT bridge is presented in "Fig. 2". It is constructed with S355J2 steel grade and C40/50 (or C35/45) concrete class for prefabricated slab of girder. The in situ plate is of C35/45 concrete class. Initially, the static scheme of a continuous beam was used and eventually a single span simply supported beam was adopted, with a span range of 15-43 m. Later on a semi-integral solution with 2, 3 and 4 spans was applied for spans in the range of 23-40 m. Nowadays the integral single span or two-span solution is considered to be a typical one for standard overpasses by VFT method. Currently the typical span for VFT bridges is about 40 meters for single span and 30meters for 2-span frames. In this span range composite construction is considered a competitive solution to concrete ones, but last year an integral bridge with a span exceeding 60m was applied in Poland. The typical solution for Poland is the fabrication of prefabricated concrete slabs at the construction site at ground level. The other aspects of structural details and fabrication are very similar to the ones used in Germany, so they are omitted in this paper.

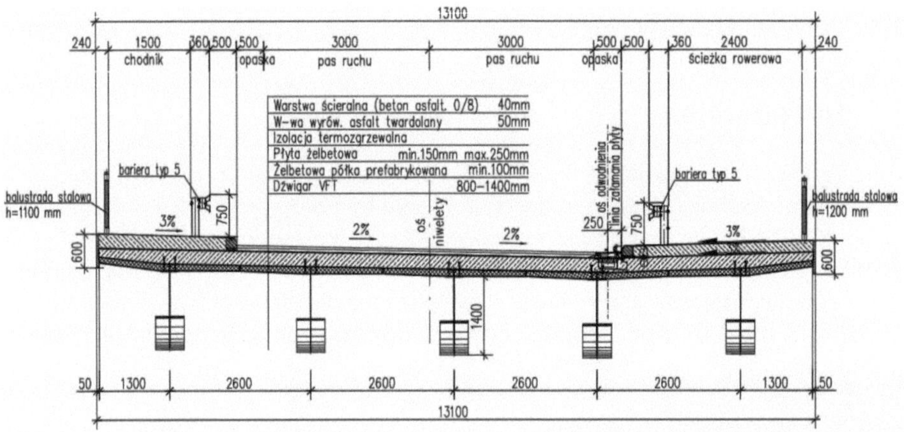

Figure 2:   Typical cross-section of superstructure of VFT single-span frame bridge in Poland

To get an impression of how the VFT method is used in Poland the last examples of the most typical integral solutions are presented in later paragraphs and they are supplemented with some specific examples of structures using rolled girders and unusual technology of fabrication.

2.1   Example of single span integral solution

The typical solution for a single span frame is presented in "Fig. 3". This is the road viaduct above the bypass of Gołdap (national road no. 65). The span is usually about 40 meters. The foundation is usually by two rows of piles or by single row of piles if possible. The superstructure is characterized by high slenderness, its geometry is presented in "Fig. 4" and "Fig. 5".

An additional coverplate under the bottom flange at mid-span is usually used if big slenderness is the case ("Fig 6").

Figure 3: Typical single span frame (VFT viaduct above bypass of Gołdap)

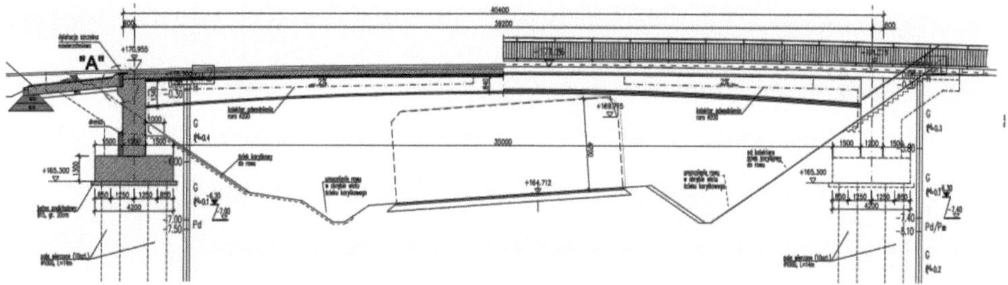

Figure 4: Longitudinal section of typical single span frame (VFT viaduct above bypass of Gołdap)

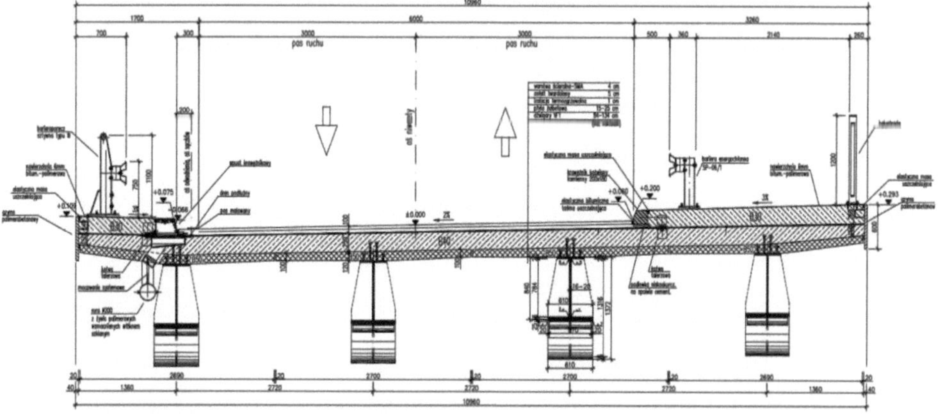

Figure 5: Longitudinal section of typical single span frame (VFT viaduct above bypass of Gołdap)

Figure 6:   Bottom view of typical single span frame (VFT viaduct above bypass of Gołdap)

A large, single span frame with 63 m span was constructed on the Route Sucharskiego (Gdańsk) for two lanes of road. The longitudinal section is presented in "Fig. 7" and the cross section of single superstructure is presented in "Fig. 8".

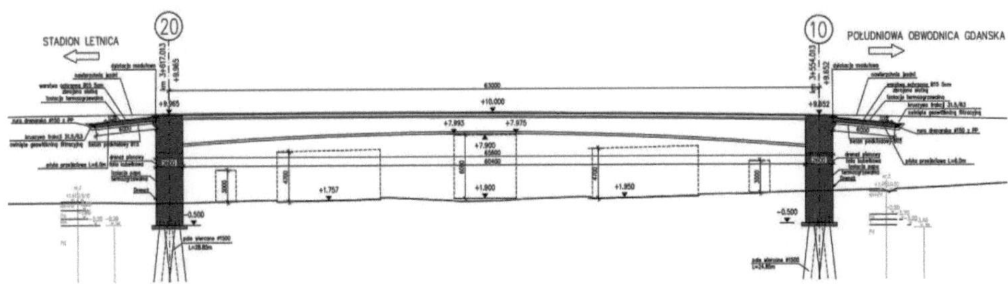

Figure 7:   Longitudinal section of 63m frame on Route Sucharskiego

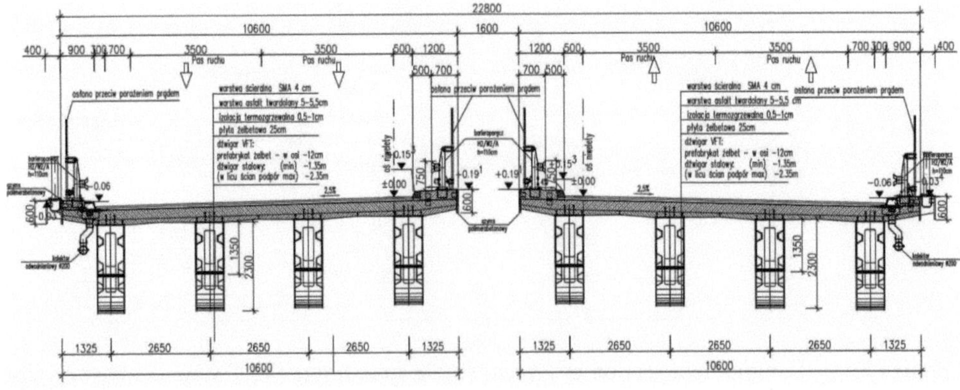

Figure 8:   Cross-section of 63m frame on Route Sucharskiego

Contrary to the standard solution (open steel section without ribs), boxed girders with longitudinal ribs were used in this case. The abutment thickness was 2,60 m. Finally one row of foundation piles was substituted by solution with two rows of piles at the construction stage. This is the largest span of a VFT bridge in Poland (and maybe the biggest in general).

Figure 9:   Using two cranes for lifting of a prefabricated girder of 63 m frame on Route Sucharskiego

Figure 10:   Placement of prefabricated girder of 63 m frame on Route Sucharskiego

Figure 11:   Construction of frame corner of 63 m frame on Route Sucharskiego

2.2   *Example of 2-span integral solution*

This solution (Fig. 12) was set to be typical for the overpass above new road (two lanes) S7 at Olsztynek – Nidzica sector from km 175+800 to km 203+600 and for the bypass of Olsztynek (national road no. 51) from km 109+500 to km 115+500 realized by design-and-build formula in the years 2009-2012. There were 17 such bridges realized and finished in 2012 with different configurations concerning the skew angle, the arching and the span lengths (approximately 2x30 m). However two span structures with fixed connection at the pillar are common solutions in Poland, this type of fully integral system with two spans was not so common for bridges with length of about 70 meters.

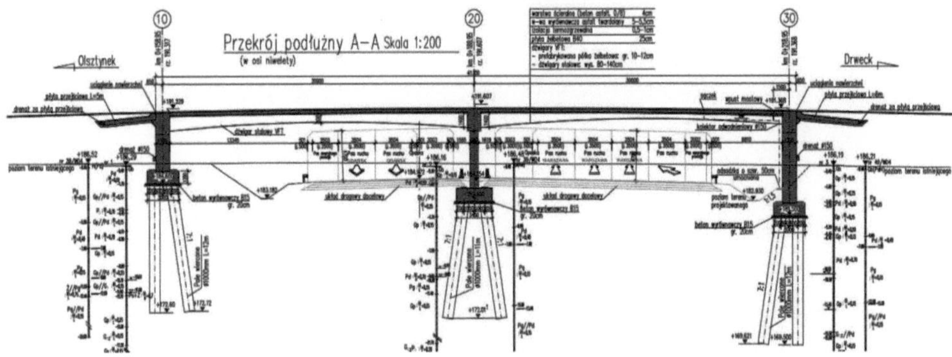

Figure 12:   Longitudinal section of typical 2-span frame above S7 road

Figure 13:   Typical solution of overpass at S7 road – 2 span VFT integral bridge

Due to the large amount of this type of bridges and good conditions concerning transportation of girders, all prefabricated VFT girders from this contract were realized in prefabrication halls and transported by trucks to the construction site, which is not so common in Poland ("Fig. 14").

The concept of constant thickness of bottom flange with appropriate shaping of height of girders was used ("Fig. 15").

Figure 14:   Montage of typical solution of overpass at S7 road – 2 span VFT integral bridge – directly from trucks

Figure 15:   Additional scaffolding of external beam and constant thickness of bottom flange is visible.

Figure 16:   Connection of steel girders at the pillar

Figure 17:   Connecting bars and reinforcement of crossbeam at the pillar

Figure 18:   Upper surface of prefabricated VFT girders

Figure 19:   19 VFT frame without finishing works and embankments

The inspection done after one year (full year season with high temperature gradients summer versus winter time) proved that there are no cracks and no evidence of significant movement of abutments can be noticed for realized bridges. This was the case for different configurations including significant skew. The solution applied for this critical detail ("Fig. 20") consisted of a special layer of composite applied in the asphalt and two chases 20 mm width spaced about 400 mm filled with bituminous material. Extensive documentation of the construction of VFT and VFT-WIB bridges at the S7 road and many pictures can be found at the web page: http://s7.olsztynek-nidzica.zaprojektuj-zbuduj.pl/roboty-mostowe.html.

Figure 20:   Asphalt layer at the abutment region (different skews)

2.3   *Example of semi-integral solution*

The semi-integral solution for a 165 m long bridge ("Fig. 21") was applied in 2012 for the construction of the bridge in Kluszkowce over Kluszkowianka river at km. 19+782,00 of the road DW 969 (sector Nowy Targ – Nowy Sącz). The system consists of 5 spans (lt = 25 m + 36 m + 48 m + 36 m + 20 m).

Figure 21:   General view of the bridge in Kluszkowce

The designed longitudinal section of the bridge is presented in "Fig 22". Contrary to usually applied solutions for integral bridges, the bottom flange is not curved but constant construction height is applied in fields and it is increased only next to middle supports. Due to the large slenderness of the web in this region (because of significant construction height) longitudinal stiffeners are applied next to supports.

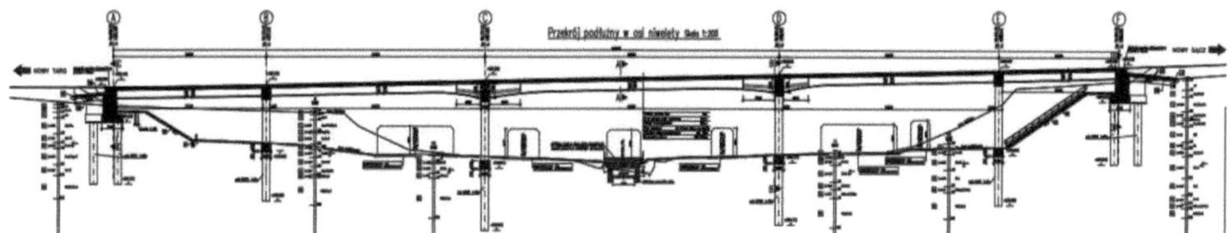

Figure 22: Longitudinal section of the bridge in Kluszkowce

The mid-span cross section is presented in "Fig. 23" and the cross-section at the pillar is presented in "Fig. 24".

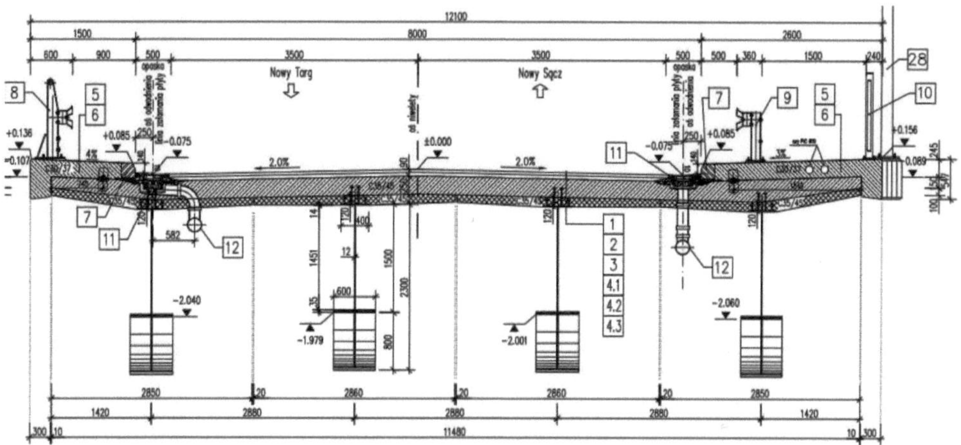

Figure 23: Mid-span section of the bridge in Kluszkowce

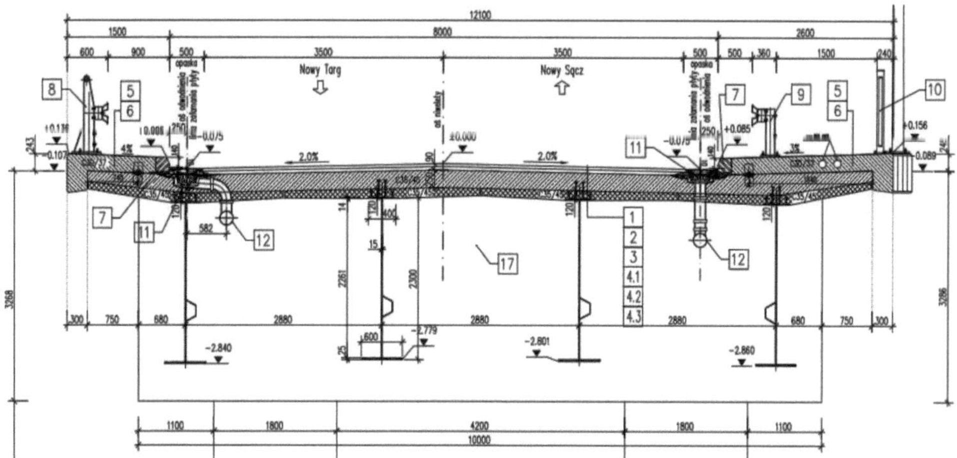

Figure 24: Cross section at the pillar of the bridge in Kluszkowce

Figure 25: General bottom view of the bridge in Kluszkowce

Figure 26: Abutment and first pillar of the bridge in Kluszkowce

2.4   VFT by rolled sections

Recently rolled sections of high strength steel (S460) become competitive solutions comparing to ordinary steel welded girders out of normal steel (S355) and it influences the VFT technology also. Road bridges WD7 and WD8 on Route Sucharskiego (Gdańsk) were realized using prefabricated composite beams with high-strength hot rolled sections after the year 2010. Viaduct WD8 is an about 390 m long continuous beam with 40 m spans ("Fig. 27"), viaduct WD7 is shorter.

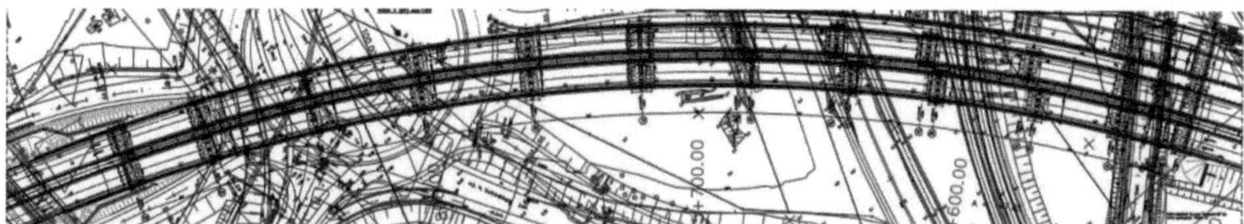

Figure 27:   Plan view of WD8 bridge.

In both cases the initially designed solution by VTF with plated girders with variable constructions height ("Fig. 28") was re-designed to rolled sections HL1100 with constant construction height ("Fig. 29"). This was the general contractor's wish at the contract stage, consequently the design office worked out such an alternative solution.

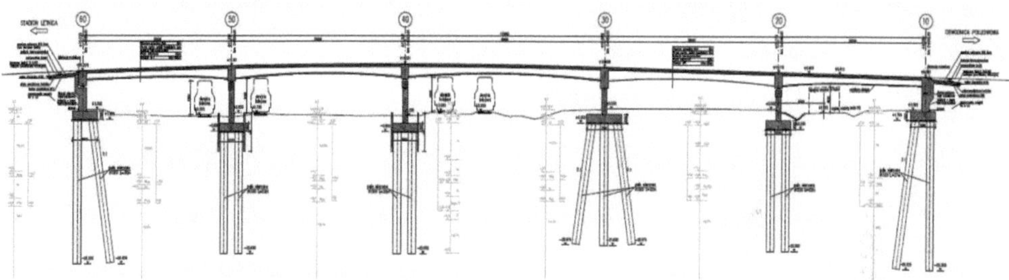

Figure 28:   Longitudinal section of WD7

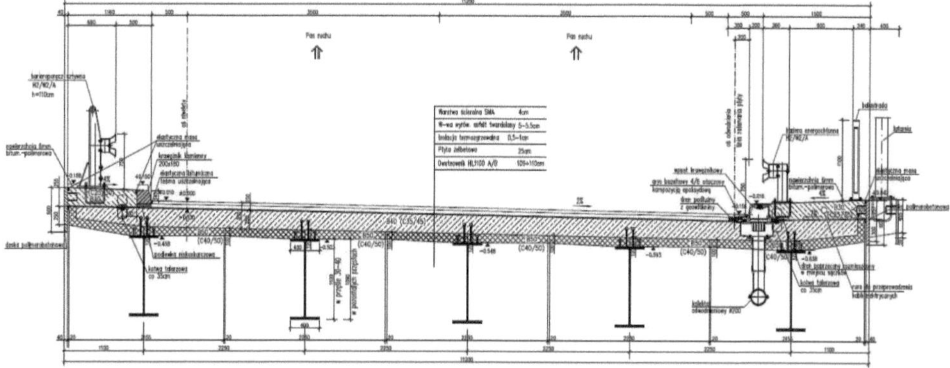

Figure 29:   Cross section of WD7 with VFT® with HL sections

Starting in June 2011, the 137 beams were rolled to final lengths between (25 m and 41 m) at the mill of Differdange, Luxembourg. All beams belong to the slender HL1100 family (steel: HISTAR460), which was developed for bridge applications. Afterwards the beams were cambered and the necessary beam finishing (welding of reinforcing plates below lower flanges at intermediate piers, end plates, headed studs and painting) was done directly on the steel plant. Thanks to this effective steelwork purchasing, the first fully finished steel girders for the WD7 viaduct were delivered in Gdansk already in August 2011 (by train up to Gdansk, and by truck to the construction site). Due to the relevant size of the bridges, the contractor decided to pour the prefabricated concrete slab in a yard near to the final site. Maximal element weight could reach 50 tons.

Figure 30:   Steel beam for WD8 girders

Figure 31:   VFT prefabricates for WD8

Figure 32:   Setting of prefabricated beam at WD8

Figure 33:   WD8 construction site

Figure 34:   View of the decks of realized viaducts WD7 and WD8, Gdańsk, Poland

2.5   *Atypical technology of erection*

Complicated and unusual technology for the construction of a VFT bridge was applied in 2008 for the bridge over Vistula River next to Cracow at highway A4 km 417+888. It was applied because of the weir under the bridge. The existing solution by simply supported concrete girders was substituted by continuous VFT beam. The technology assumed removal of existing spans and implementation of new ones span-by-span, using two cranes standing at the edge of spans neighboring the replacing one. For this reason a special steel platform and modification of the VFT girder construction were applied so they could handle very heavy load from the cranes moving old 80 ton girders and placing new 40 ton prefabricates. It enables in addition parallel crane operation on longitudinal launched pedestrian bridge.

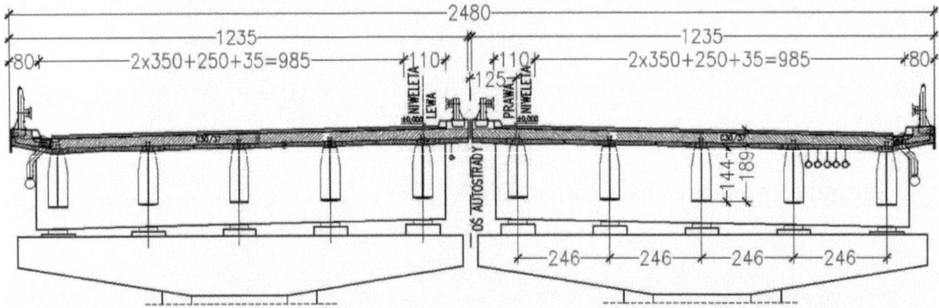

Figure 35:   Cross-section of new the VFT bridge at highway A4 km 417+888

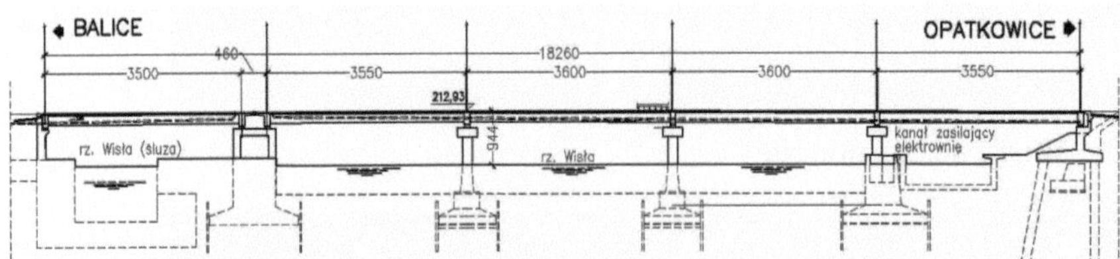

Figure 36:   Longitudinal section of new VFT bridge at highway A4 km 417+888

Figure 37:   Old concrete span removed at the bridge at highway A4 km 417+888

Figure 38:   Crane on temporary platform.

Figure 39:   Modification of VFT girder for implementation of temporary steel platform

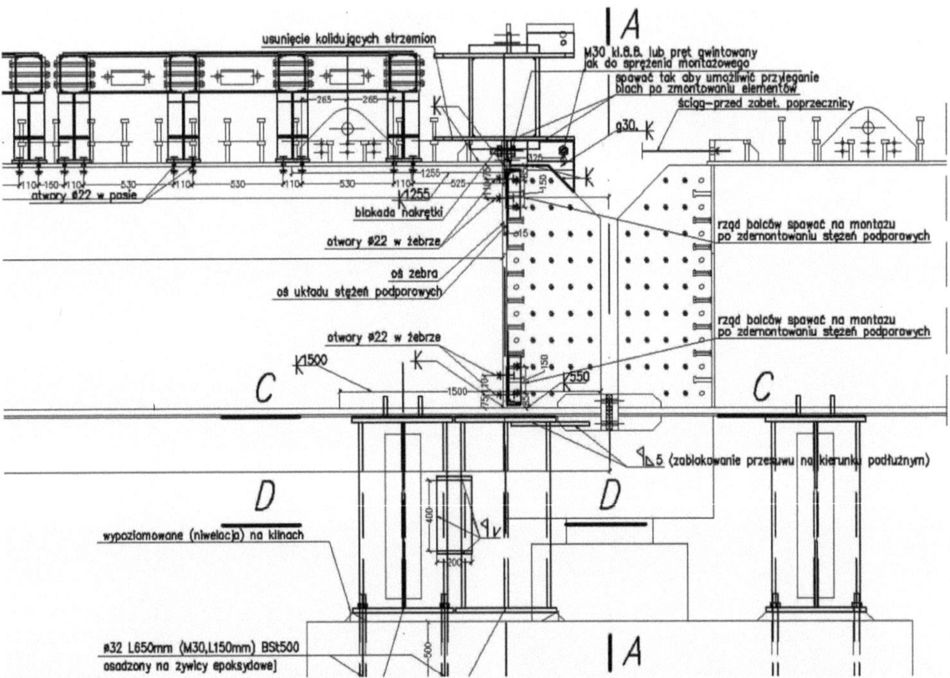

Figure 40: Strong temporary support of steel girders and connection with steel platform

Figure 41: Special steel structure and modification of ordinary VFT solution of steel girders for temporary support of crane

Figure 42: The crane on prefabricated VFT girders operating on the pillar of neighboring pedestrian bridge under construction

Figure 43: Realized first line of VFT continuous beam (from right)

Figure 44:   Montage of prefabricated elements

Finally the first lane of the bridge was realized with success by such complicated technology and the second was realized using montage from ships just because free access from the water was available in that case.

3.Conclusions

Recent bridges by VFT method in Poland are presented in the paper and trends are highlighted on the background of solutions being used during the last decade. The system is very flexible concerning possible static schemes and the fabrication technology of girders. One can notice that the most common solution for Polish VFT bridges is construction of prefabricated elements next to the construction site, but an example of S7 road with 17 bridges being realized in different way proves, that it can change towards the solution that is used in Germany (full prefabrication and transport of composite prefabricates on site). The VFT technology is used in Poland for road bridges nowadays because not so many contracts for railway bridges appeared in the last years to be suitable for VFT method. However, the implementation of VFT-WIB technology for two railway bridges in Poland (including the first implementation in the world) proves that in future the VFT method will be probably applied for railway bridges also.

Renewal of old existing small road bridges with modular system – CASE STUDY Mânărău BRIDGE

Edward Petzek[1], Luiza Toma[2], Elena Meteş[3] & Radu Băncilă[4]

Keywords: EcoBridge, VFT-WIB®, precast composite slab, integral bridge, composite dowel, modular solutions.

Abstract: Romania has a highway network of approx. 154 000 km on which there are over 3300 bridges. The usual range of these bridges is situated in the field of medium spans and was dominated in the past by concrete bridges. At the same time more than 50 % of the total number of bridges is designed for the loading class I (Lorries A 13 tones, S 60 tonnes). For them the National Administration of Roads proposes (in a long term program – Romanian Highway Administration Report 2003) their rehabilitation and upgrading to Loading class E (Lorries A 30 tones, V 80 tonnes). The European RFCS project ECOBRIDGE *or Demonstration of ECOnomical BRIDGE solutions based on innovative composite dowels and integrated abutments* [1] started in 2010 and has the principal objective to continue the research, demonstration projects and accompanying measures for the promotion of knowledge gained by the two pilot projects: INTAB – "Economic and Durable Design of Composite Bridges with Integral Abutments", 2005 – 2008 and PRECOBEAM – "Prefabricated Enduring Composite Beams based on Innovative Shear Transmission", 2006 – 2009. The main conclusion of the two pilot projects was that the composite dowel strips combined with the integral structure concept find a wide application in bridges. The article presents the first Romanian proposed project for the European RFCS program in which the composite dowel strips together with integral abutments in terms of construction details and economic aspects are implemented.

1.   Introduction

Rehabilitation and maintenance of existing steel constructions, especially steel bridges, is one of the most important present problems. During service, bridges are subject to wear. In the last decades the initial volume of traffic has increased. Therefore many bridges require a

1   Assoc.Prof.Dr.Ing., Politehnica University of Timşoara and SSF-RO Ltd. Spl. T. Vladimirescu 12/6, Timişoara, Romania, epetzek@ssf.ro
2   Drd.Ing. Politehnica University of Timşoara and SSF-RO Ltd. Spl. T. Vladimirescu 12/6, Timişoara, Romania, ltoma@ssf.ro
3   Drd.Ing. Politehnica University of Timşoara and SSF-RO Ltd. Spl. T. Vladimirescu 12/6, Timişoara, Romania, emetes@ssf.ro
4   Prof.Dr.Ing., Politehnica University of Timşoara and SSF-RO Ltd. Spl. T. Vladimirescu 12/6, Timişoara, Romania, radu.bancila@ct.upt.ro

detailed investigation and control. The examination should consider the age of the bridge and all repairs, the extent and location of any defects etc.

The research program of the European RFCS project ECOBRIDGE was split in three working groups: one partner was from Germany (Aachen University, SSF Ingenieure AG, TWT Sanierungsgesellschaft), another one from Poland (Wroclaw University, Europrojekt Gdansk and Energopol Szczecin companies) and one from Romania ("Politehnica" University from Timisoara, SSF-RO s.r.l., D.R.D.P. Timişoara). Arcelormittal coordinates the activities of all working groups.

The propose of the RFCS research project ECOBRIDGE was to design, construct and have a monitoring period for three composite bridges in Romania, Germany and Poland with integral abutments and / or composite dowels – an innovative form of shear transmission. The main objective of the first task was to identify a proper bridge site by considering socio-economic as well as technical conditions.

At the beginning, Romania has chosen for this European project an existing bridge at Mânărău. During service, the integrity of the bridges can be influenced by many factors like the increase of the initial traffic volume, the existence of an inadequate maintenance process or a total lack of it, natural disasters, the oldest of them could even be subject of world wars. The imprint of those factors on a structure can be pointed on the appearance of fatigue defects, contingent deformations, corrosion levels, behaviour due to traffic loads or the bearing condition. All these aspects are valid for the Mânărău Bridge, chosen to be deconstructed [2].

The answer to the question "why this structure?" is the following: this structure has been chosen because it was established that is in a very bad condition and doesn't offer safety in operation anymore and also because many other similar small bridges, built in the same period of time are still in use with relevant problems as well.

Moreover, the existing bridge was proposed to be replaced and there has been a technical project that included a classical filler beam deck solution. The technical proposal stipulated for the renewal works to be executed on the initial location of the bridge, while the continuity of the traffic was to be assured during the entire working period by setting up a bypass with a single traffic lane and traffic lights. This temporary traffic variant implied the introduction of a temporary steel bridge with a span of 15 m. The total length of the variant was to be of 130 m and the speed limit would have been of 15 km/h.

Taking these aspects into account, within the applied research programme a modular structure was conceived, which could have been executed on one traffic lane at once under the protection of small longitudinal supporting structures, thus eliminating the costs for the traffic variant. At the same time, the settled objectives were to also minimize the investment costs whilst keeping and assuring the structural robustness. Thus, by using compound dowels it was possible to reduce the steel quantity in the external rigid reinforcement embedded in the concrete and by using the solution of an integral bridge on abutments made of sheet piles one could guarantee a 100% modular structure, which could be easily and rapidly executed.

The bridge is situated on the National Highway DN 79A Km 60+627, near to the village Mânărău in the Arad County ("Fig. 1").

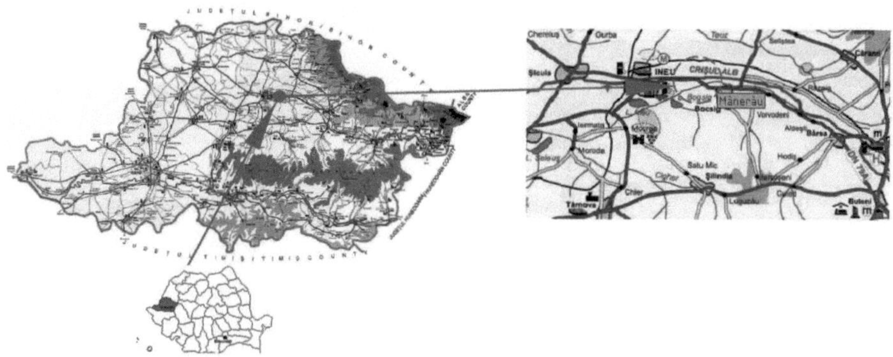

Figure 1:   The Mânărău bridge location

Table 1:   Technical information of the existing vs. the new bridge

	The new structure	The existing structure
Importance category (HG 766-97)	C	C
Category of construction (STAS 4273-83 art. 2.11)	3	3
Class load	E (A30, V80) and LM1	I (A13, S60);
Bridge length	8,175 m;	9,85 m
Bridge width for the solution 1	10,36 m;	9,00 m
Bridge width for the solution 2	10,08 m;	
Gradient's bridge	0 %;	0 %;
Gradient in the cross section	2,5 %;	2,5 %;
Length of the bridge guardrail	32,2 m	0,0 m
Connection with embankments	connection plates, back walls	back walls
Static structure	frame structure with an opening	simply supported girder
Waterproofing	waterproofing membrane with a protection layer	waterproofing membrane with a protection layer
Access ramps	rehabilitation on 2 x 20 m and will be referred with road verges of 0.5 m;	-
Infrastructure	indirect foundation	direct foundation
Drain water from the bridge	side ditch	-

Figure 2:   General view of the bridge and access ramps

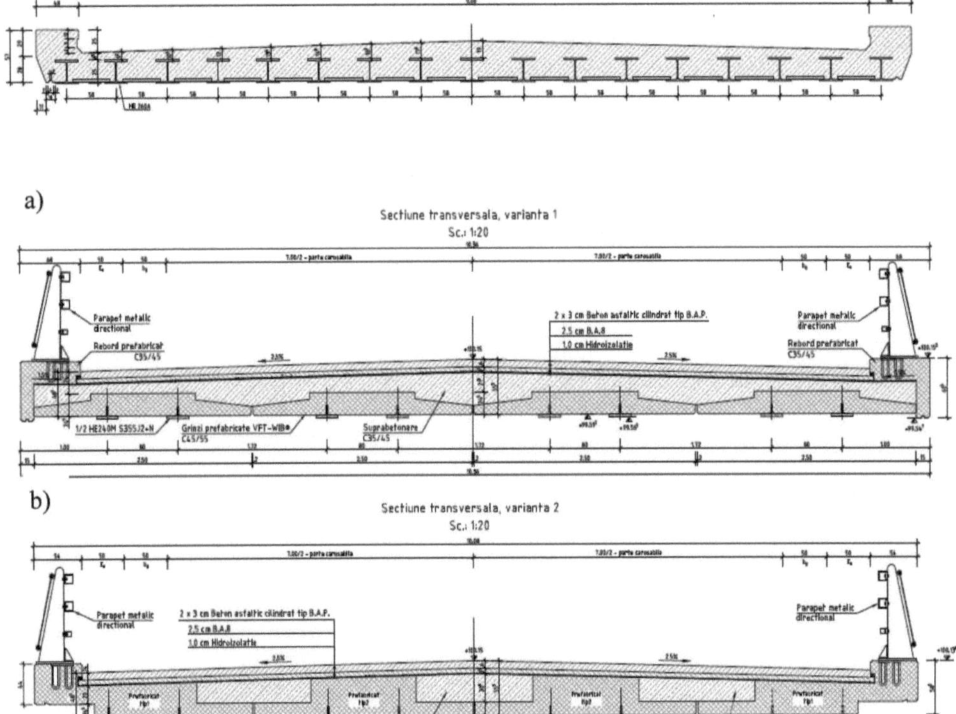

Figure 3:   General cross section: Old solution, new solution 1 (a), new solution 2 (b)

The bridge was built in 1967 for the loading class 1 (truck convoys A13 and S60). The general condition of the structure is bad; the maintenance is missing ("Fig. 2").

The bridge belongs to the Regional Administrations of Roads and Bridges (DRDP). It does not correspond anymore to the present traffic necessities, with trucks of 30 tons (A 30)

or even more. There are no footways and no borders. The safety parapet is missing, the traffic participants being in danger. Water discharging devices are also missing on the bridge.

The total length of the bridge is 9,90 m and the width is 9,0 m. The present cross section consists of a reinforced concrete slab of C8/10, having a thickness of approx. 0,40 m ("Fig. 3"). The carriageway, made out of asphalt concrete, presents cracks on extended areas. The infrastructure presents degradations, caused by waters ("Fig. 4"). There is a geotechnical study based on geotechnical investigations, presenting the layers of the foundation ground. The bridge is situated in a seismic zone; according to the Romanian Standards, no measures for anti-seismic protection have to be taken.

Figure 4:   Structure degradations

The present viability state of the structure is not satisfactory, which leads to the necessity of replacement with a new structure. The highway bridge presents damages as a result of actions, fatigue and creep. The replacement of the existing structure with the VFT-WIB® solution (developed from the classical WIB composite structure) was proposed.

2.   Technical details of the applied solutions

A classical filler beams deck could be an adequate solution for the span and heights imposed for many existing structures especially railway bridges. Going further and taking into consideration also the need of a simple technology and a very short erection time, it leads to the necessity of a modular system with low costs. Using the high degree of prefabrication the possibility of unexpected situations on site is reduced and lower costs are obtained. Simultaneously it offers execution simplicity. The VFT-WIB® solution was obtained from the classical WIB composite structure, considering the improvements mentioned before ("Fig. 5").

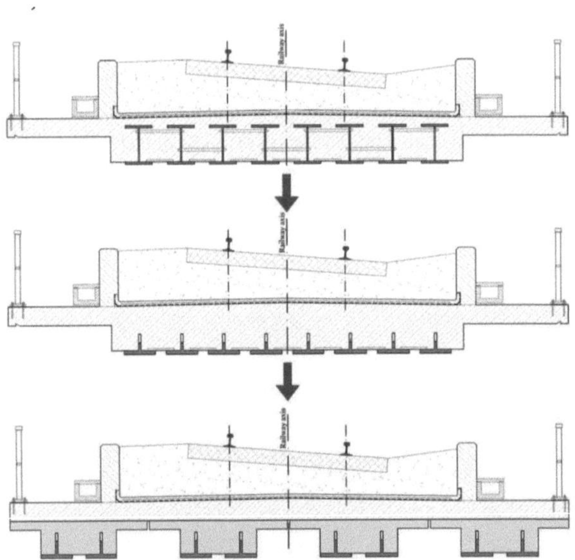

Figure 5:   Durability of composite bridges

This new technology has been in continuous development in Europe.

The composite dowels are associating steel T-sections acting as tension member with a concrete top chord acting as compression member. Steel parts are generally obtained from rolled steel profiles that are longitudinally cut in two identical T-sections. The cut is performed with a special shape to allow the shear longitudinal transmission between steel and reinforced concrete ("Fig. 6" [4]). Also a good accuracy of the cutting line is necessary because an imperfection in the geometry can compromise the final resistance to fatigue [5].

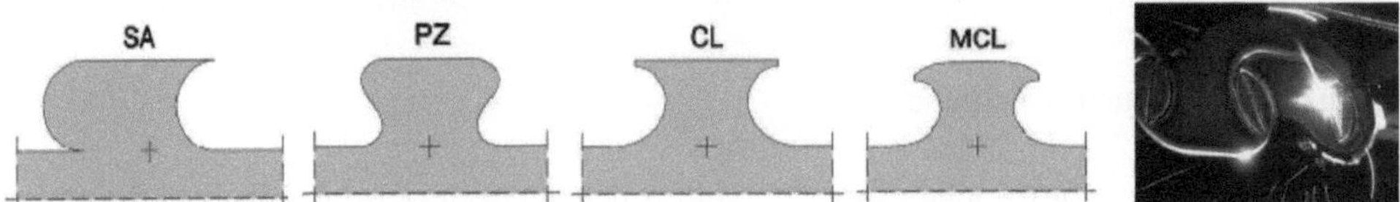

Figure 6:   Cutting line types: fin (SA), puzzle (PZ), clothoidal (CL), modified clothoidal (MCL).

Going further by introducing these beams into framed load bearing systems such as integral abutment bridges, it is possible to manufacture hybrid integral structures with concrete shafts and composite crossbeams that can easily absorb the high horizontal forces from impacts or earthquakes. In case of an integral system, due to the absence of bearings and joints, the maintenance costs can be significantly decreased [3]. Also the reduction of the number of construction stages on the site leads to a simple technology and a high quality assurance, during the building process. Those main advantages of the integral bridges turn out to become highly attractive to designers, constructors and road administrations.

Over the last four years, VFT® and VFT-WIB® technology included within frame bridges has been applied in Romania. Unfortunately, as there occurred the same financial problems in case of the bridge over the Mânărău channel on the national road, all the works have been stopped and the project remained only in design stage.

3. Design aspects

With the help of the main dimensions and cross sections of the structural elements from the existing drawings and completed by the present situation on the site, a simple analysis of the structure with the help of a FEM analyses was made. According to the results, the main girders subjected to current Romanian standards exceed the normal values. The conclusion was that the existing bridge needs to be replaced. The new structure was developed and also analysed in a FEM program according to the Eurocode load models ("Fig. 7").

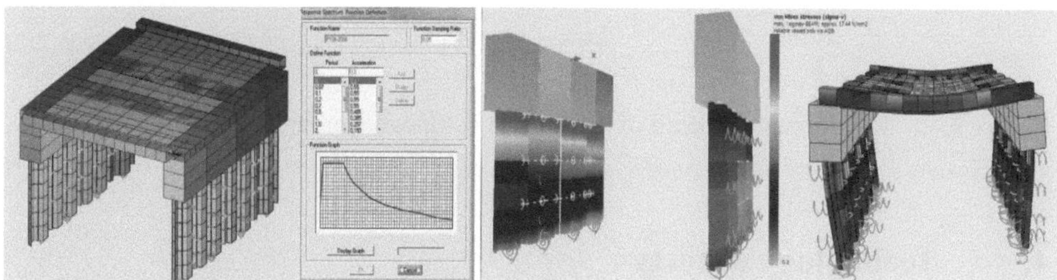

Figure 7: Spatial 3D model and FEM analysis

Therefore, for the new cross section two types of precast elements were proposed. The difference between the two is only the geometry of the elements ("Fig. 8"). Four VFT- WIB® prefabricated composite girders were aligned and linked together by cast-in-place areas. To reduce the usage of the formwork on site to the maximum, precast concrete solutions were also provided for the sidewalks ("Fig. 8").

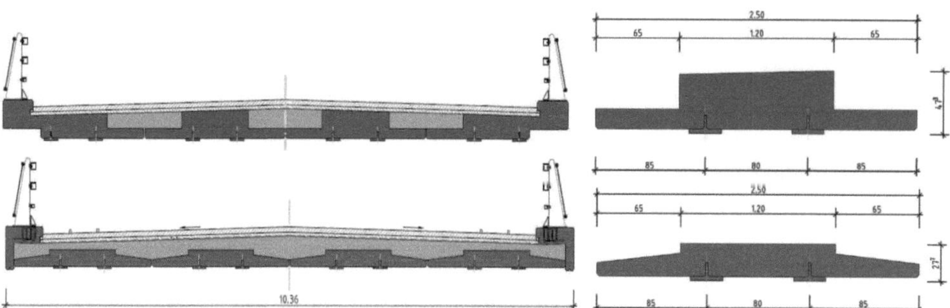

Figure 8: General new cross sections and details with the precast elements

Considering only one precast element, the steel girders should be composed of two rolled girders ½ HE240M at the bottom of the section, made of steel S355 J3+N. A rolled steel girder is cut according to a separation line (corresponding to the modified clothoidal or MCL shape) [3], resulting 2 "T"-shaped steel girders. The resulted halved girders work as external reinforcement, the steel consumption is reduced to a minimum, leading to a very slender and economical composite structure. The MCL composite dowels are suitable for the bridges field, because they allow high bearing capacities and provide uniform and bidirectional transmission of the shear forces in the structure, and can assume also the dynamic loads. They have a good behaviour also in longitudinal direction.

The reinforcement bars are passing perpendicularly to the web of steel profile and through the concrete area between the steel dowels. Achieving its role in the composite dowel, it must also resist to the shear forces ("Fig. 9").

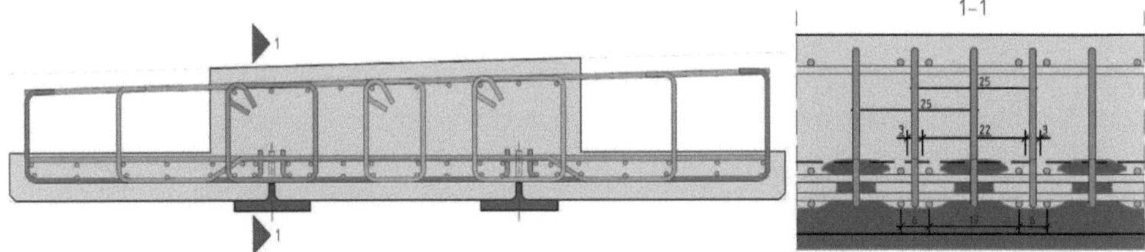

Figure 9: Reinforcement detail of the precast composite girder

4. Technology aspects

Integral bridges require full collaboration between structure and foundation soil. The solution with Larssen sheets disposed in the bearing axis was adopted; they transmit the loads of the superstructure in the carrying soil foundation ("Fig. 10").

The following technological phases have been proposed:

- The Larssen profile (type 604) having a total length of 9,0 m will be introduced in the soil.
- At the top of the Larssen profiles a bearing seat of reinforced concrete C25/30 with a width of 1,00 m will be provided.
- The abutments will have back walls and connection plates of reinforced concrete C25/30.
- 4 rolled steel girders HE240M of S355 J3+N – resulting 8 steel "T" shaped beams will be prepared; 2 for each precast element – as a rigid external reinforcement.
- Bst500 reinforcement, C45/55 concrete – for the precast elements.
- After 14 days the prefabricated girders will be transported on site and placed in final position. Due to the high degree of prefabrication the influence of the shrinkage and the creep on the structure is eliminated.
- In this phase the structural system is a simply supported girder.
- Finally the precast beams are fixed at the ends with concrete class C35/45 obtaining the frame effect, resulting a frame system ("Fig. 10").

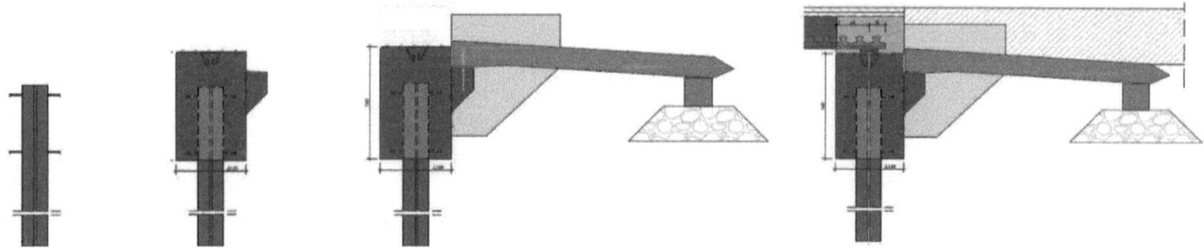

Figure 10:   Frame corner technological phases.

5.   Conclusions

Combining the advantages of the VFT® girder with the robustness of the traditional "filler beam plate" a very slender, robust, durable and economical structure was obtained due to an optimal usage of materials and short construction times. Due to the absence of bearings and joints, the maintenance costs were also significantly decreased.

By eliminating the traffic variant which included the temporary bridge, the investment costs as well as the execution time could be substantially reduced.

References

[1]   RFCS research project, "ECOBRIDGE: Demonstration of economical bridge solutions based on innovative composite dowels and integrated abutment, Grant no. RFSP-CT-2010-00024", 2010

[2]   Seidl, G, Petzek, E, Băncilă, R, "Composite Dowels in Bridges – Efficient solution", Timişoara, Trans Tech Publications, Advanced Materials Research Vol. 814, ISBN-13: 978-3-03785-848-6, 2013

[3]   Seidl, G, et al., "Prefabricated enduring composite beams based on innovative shear transmission – Final Report RFSR-CT-2006-00030", s.l.: European Commission, 2011.

[4]   Berthellemy, J, Lorenc, W, Mensinger, M, Rauscher, S, Seidl, G, "Zum Tragverhalten von Verbunddübeln – Teil 1: Tragverhalten unter statischer Belastung", Stahlbau 80, Heft 3 2011, 172-184, 2011

[5]   Barthellemy, J, Hechler, O, Lorenc, W, Seidl, G, Viefhues, E, "Premiers resultants du projet de recherché Europeen PreCoBeam de connexion par decoupe d'une tole", Bulletin Ouvrqges d'Art du CTOA du Setra, n°3, 2009

Integral bridge using the VFT-WIB technology for a three-spanned structure

Edward Petzek[1], Elena Meteş[2], Luiza Toma[3] & Radu Băncilă[4]

Keywords: VFT-WIB, prefabricated composite beam, composite dowel, integral bridge, economical bridge solutions.

Abstract: Alongside Romania's main focus direction of the past years regarding the transportation system, the motorways, which are part of the Pan-European corridors, the entire road network has to be considered. This is the case of the local road which crosses over the Bistriţa River, in the village Frânceşti, via a provisory bridge, used as permanent solution. The bridge does not ensure the safety of its passengers, nor the required clearance. The conducted expertise concluded that the structure must be replaced. The new bridge to be built was chosen to be part of the European research project ECOBRIDGE [1]. It is an three-spanned integral bridge using the VFT-WIB® technology for its superstructure. This technology referring to the German "Verbund-Fertigteil-Träger – Walzträger im Beton" or "prefabricated composite beam – filler beam" was introduced to the market in the last years. A key feature of the solution is the use of composite dowels for the steel-concrete connection. Important European research programs such as INTAB [2] or PRECOBEAM [3] have shown the use of such dowels is efficient and that VFT-WIB® is suitable to be used in case of integral bridges, combing the advantages of both technologies.

1.   Introduction

The European research project *Demonstration of ECOnomical BRIDGE solutions based on innovative composite dowels and integrated abutments* with the short name ECOBRIDGE [1] has the objective to construct three composite bridges in Romania, Germany and Poland with integral abutments and/ or composite dowels – an innovative form of shear transmission. This is a possibility to apply the newest techniques and developments in each participating country. The bridge at Frânceşti was the second proposal, made by the project members from Romania, to be part of ECOBRIDGE.

---

1   *Assoc.Prof.Dr.Ing., Politehnica University of Timişoara and SSF-RO Ltd. Spl. T. Vladimirescu 12/6, Timişoara, Romania, epetzek@ssf.ro*

2   *Dipl.Ing., Doctoral student, Politehnica University of Timişoara and SSF-RO Ltd. Spl. T. Vladimirescu 12/6, Timişoara, Romania, emetes@ssf.ro*

3   *Drd.Ing. Politehnica University of Timişoara and SSF-RO Ltd. Spl. T. Vladimirescu 12/6, Timişoara, Romania, ltoma@ssf.ro*

4   *Prof.Dr.Ing., Politehnica University of Timişoara and SSF-RO Ltd. Spl. T. Vladimirescu 12/6, Timişoara, Romania, radu.bancila@ct.upt.ro*

The bridge is located in the village Frâncești and crosses over the Bistrița River. The existing bridge is a provisory structure ("Fig. 1"), not ensuring the safety of its passengers, with a total length of 75,00 m, having 10 spans each of approximately 7,50 m length and a width of 2,40 m. It is the only crossing possibility of the river for the local communities and in urgent need of renewal. Bridges are one of the main founds consumers and an important investment for a small community. The economical factor is important in terms of material consumption, environmental impact and structure costs. Efficient solutions have to be elaborated in order for the structures to be both durable and economic.

Figure 1:   Existing bridge over the Bistrița River at Frâncești

The replacing structure was designed as a three-spanned integral bridge with VFT-WIB® superstructure. Each span has a length of 21,10 m, with a total length of 64,50 m, the width is 5,20 m, with a carriageway of 4,00 m ("Fig. 2"). The design is based on the experience gained at the Vigaun road bridge over ÖBB track Salzburg – Wörgl at km 23,135, Germany built in 2008 [4].

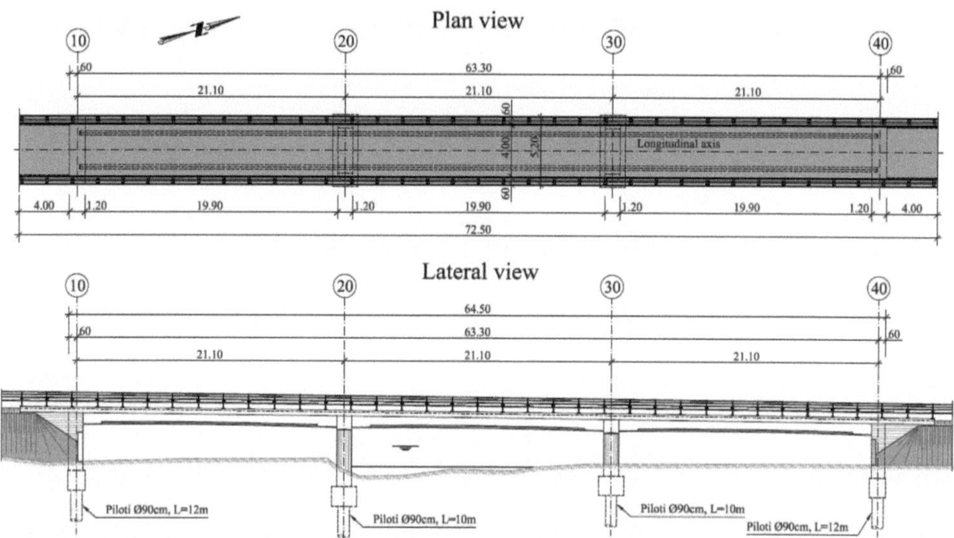

Figure 2:   Plan view and lateral view of the new bridge at Frâncești

The construction of the bridge at Frâncești began in 2012 when a large part of the infrastructure was executed. Due to lack of founds, the construction process was interrupted. At present the local administration works towards finding the necessary means to finish the structure.

The current paper presents design aspects and structural details of the VFT-WIB® structure.

1.1   Integral bridges using VFT-WIB® superstructure

Integral abutment bridges have by now demonstrated their effectiveness, especially through material consumption and maintenance savings. This type of structure requires special attention in design, considering the frame-type behavior. Modern calculus methods have made it possible for the engineers to evaluate correctly the behavior of static complex structures.

The frame effect, by ensuring smaller positive bending moments in the middle area of the spans in comparison to simple supported or continuous superstructures systems, allows for the superstructure to have a smaller constructive height and reduce the material consumption. At the same time, negative bending moments arise in the joint areas of the superstructure, imposing a strong conformation of the frame nodes, needing reinforcement at the top layers and continuity between the superstructure and infrastructure [2].

By using integral abutment bridges no expansion joints and no bearings are needed. The risks emerging from the great execution precision needed for the disposal of the expansion joint or of the bearing equipments are eliminated. Simultaneously the maintenance procedures are eased and inspection times are further apart.

The frame type structures bring about new challenges to the design as well as to the execution procedure, but do succeed in bringing advantages regarding both total costs and easiness of the montage. Efficient structures can be obtained when for integral bridges prefabricated composite girders with composite dowels are used.

The VFT-WIB® solution, a modern reinterpretation of the filler beam decks, has by now been used in countries like Germany or Poland for small and middle spanned bridge structures. Earlier research projects such as PRECOBEAM [3], focused on the design of so called composite dowels, a main feature of the VFT-WIB® solutions, prefabricated composite girders.

The standard double T rolled steel profiles are replaced by T shaped sections with a bottom flange and a web, while the concrete section is redesigned and reduced. Modularity is obtained by creating prefabricated parts. Individual prefabricated reinforced concrete beams with external reinforcement represented by the T steel girders are created. The two materials are linked together by composite shear connectors [3], with no need of headed studs and hence no need of upper flange for the steel profile. This concept allows many design opportunities and therefore many types of VFT-WIB® sections were obtained: using steel profiles „duo-WIB" ("Fig. 3a") and „mono-WIB"("Fig. 3b"), using welded sections ("Fig. 3c", "Fig. 3d"); special types VFR-Rail ("Fig. 3f") and external reinforcement in in-situ plate cross-section ("Fig. 3g") [4].

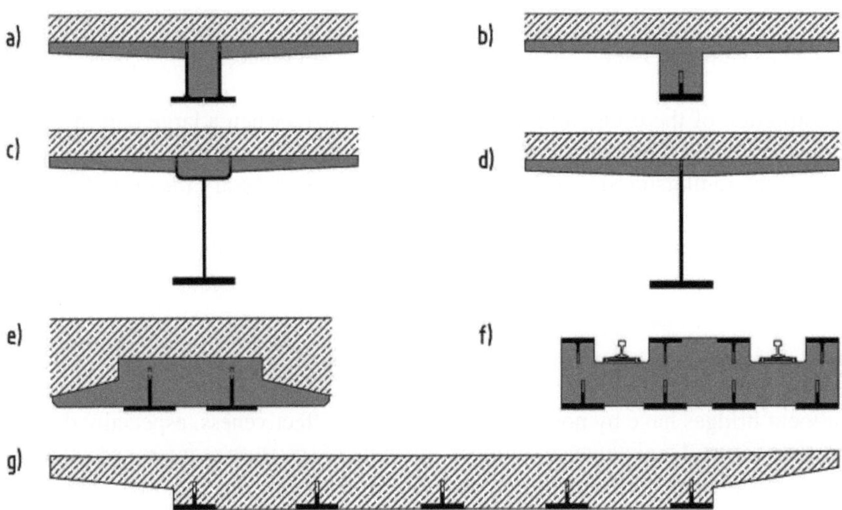

Figure 3:   VFT-WIB® section types a) to g) [4]

In case of the bridge at Frâncești the „mono-WIB" – design was chosen, suitable for spans with lengths up to 35,0 m. This VFT-WIB® solution uses one or more beam type elements displayed in the bridge's longitudinal direction, bound together by in-situ concrete decks or by joints filled with concrete and connection reinforcement. These prefabricated girders consist of an upper reinforced concrete flange, a reinforced concrete web and one imbedded T shaped steel profile at the bottom of the section using composite dowels for the shear connection. The T shaped steel sections are obtained by a special cut along the longitudinal axis of the web of a double T rolled steel profile almost without any outcuts. Two individual T profiles result, each having due to the special cut line tooth shaped steel dowels at the free end of the webs. The separation cut has to be performed precise to avoid imperfections and a possible compromise of the final fatigue resistance. The composite dowels represent the interaction between the steel dowels, through-going reinforcement bars and enwrapping concrete.

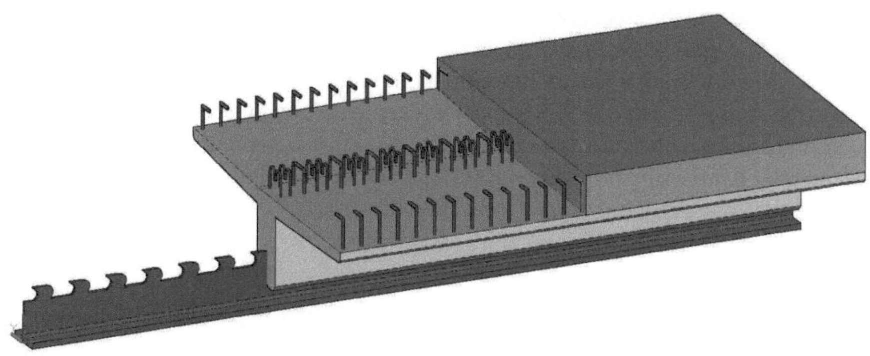

Figure 4: VFT-WIB® girder and concrete deck

The composite dowel strips were studied in the past years and design details can be found in [3], [5]. For this bridge the fin or SA shape was chosen ("Fig. 5a"), which is designed to transfer the shear forces in only one direction. Since the dowel shape is not symmetric, the dowel orientation changes at the middle of each steel girder ("Fig. 5b").

The composite dowel strip is located quite far from the neutral axis and the steel dowel not only gets local shear loads but also centric tension as a result of global bending moment and gets consequently higher fatigue loads due to global bending moments [4].

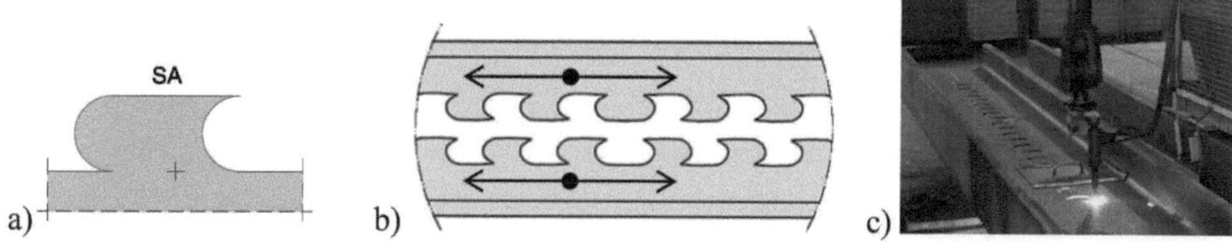

Figure 5: SA or fin shaped steel dowels [5]: The SA shape (a), Direction change of the composite dowels in the middle of the girder (b), Cut along the separation line in the plant (c)

This construction method was chosen due to advantages such as the robust bearing behavior in the case of impact loads assured by the concrete web, facile prefabrication possibilities, materials savings – especially regarding the steel use. High slenderness can be obtained if used for frame type systems.

2. Design aspects

The cross section of the bridge ("Fig. 6a") aligns two mono-WIB girders bound together by a 20 cm thick concrete deck. The design of the prefabricated girders is shown in Fig. 6b and are made each of one ½ HEM 600 steel rolled profile of quality S460 ML, and of an upper concrete flange and a concrete web of C50/60 class. The concrete flange is 10…13 cm thick and ~2,60 m wide; the concrete web is 30,5 cm wide with a height of 70 cm. For the entire three-spanned bridge with a total of six prefabricated mono-WIB girders only three rolled girders HEM 600 were used.

The ½ HEM 600 and the steel dowels are obtained by cutting a regular HEM 600 rolled profile according to a pre-established geometry ("Fig. 7", "Fig. 8", "Fig. 9") with the adequate technology in the factory. Only a low amount of outcuts result. Afterwards the girders are brought to the factory, where the dowel reinforcement, the binding reinforcement and the required carrying reinforcement are disposed ("Fig. 10"). High class concrete is poured in the formwork and after hardening, the prefabricated girders are carried to the site. The prefabricated girders can also be produced on site, near the infrastructure.

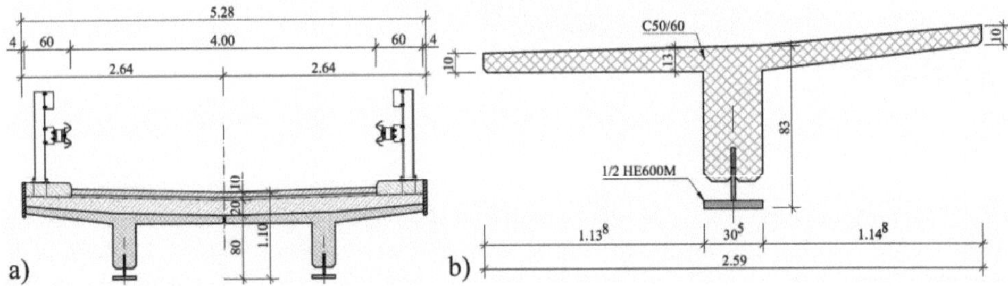

Figure 6:   VFT-WIB – mono-WIB cross section (a), General superstructure cross section (b)

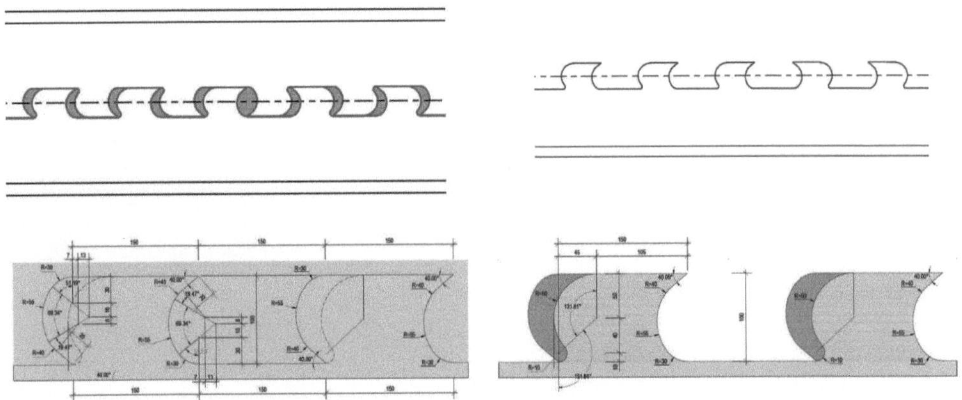

Figure 7:   Execution details of the steel girders; Geometry of the steel dowel

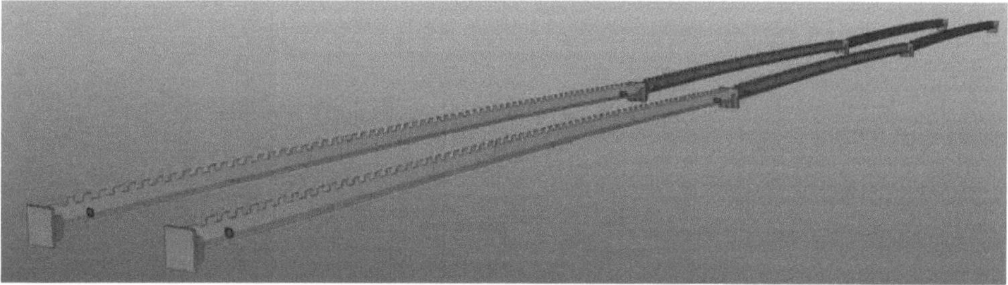

Figure 8: External reinforcement

Figure 9: External reinforcement – end detail

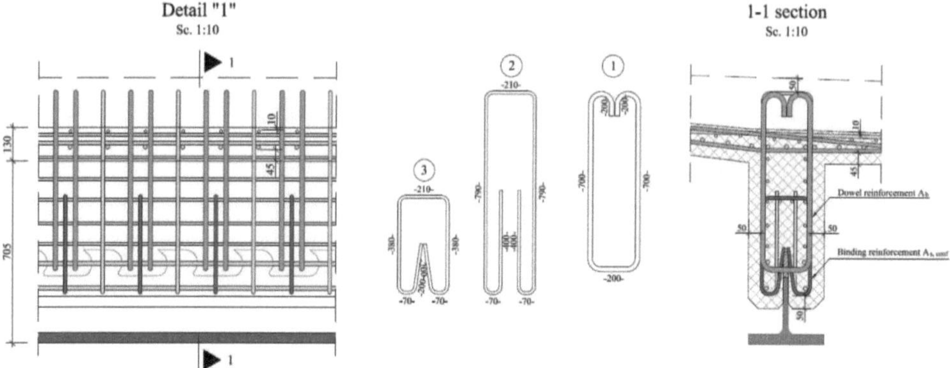

Figure 10: Composite dowels – reinforcement details

The infrastructure and superstructure of the bridge are monolithically bound together by in situ concrete and connection reinforcement from the piers/ abutments, the VFT-WIB girders, the deck and the frame nodes. The simple, rectangular shaped abutments are provided with relative small back walls and transition slab. The elevations rest indirect foundations with piles of 0,90 m in diameter. The bridge's static system ("Fig. 11") becomes a multi-span frame, resulting an integral bridge.

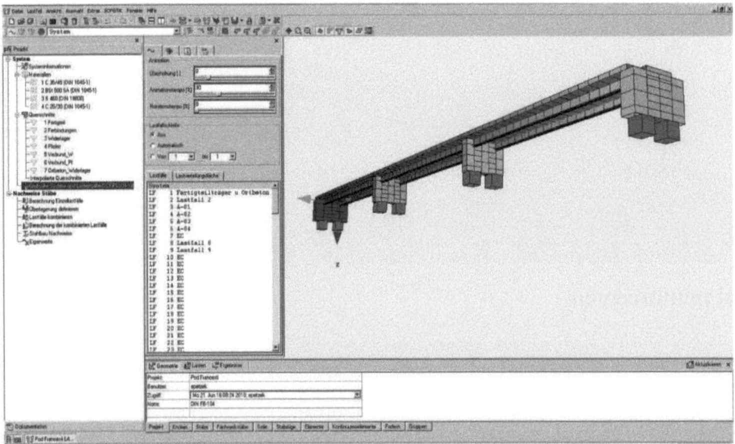

Figure 11:   Static model

2.1   Construction stages

The infrastructure consisting of piles, foundation plates and elevations is classically erected, but due to the regular, simple and slender shapes of each part is economical, easy and fast to build. The VFT-WIB girders are made either in the prefabrication workshop or in situ on small concrete platforms. After the prefabricated mono-WIB girders are brought on site, they are lifted in their final position and fixed on top of the piers and/ or abutments ("Fig. 12a"). Reinforcement bars are added at the frame nodes and on the deck, and linked to the connection reinforcement from the elevations and the prefabricated girders. After the concreting of the frame nodes an intermediary static frame system is created ("Fig. 12b"), than the concrete deck can be poured ("Fig. 12c").

The integral building method implies the connection between all carrying elements. This is realized by outgoing reinforcement from every previously built/ added part. No props are needed during the concrete casting and only lateral formworks for the in situ deck are used due to the "T" -shaped VFT-WIB sections. The weight of the bridge's deck fresh concrete is taken over from the frame system: the two composite prefabricated girders, the frame nodes and the infrastructures. An optimal total superstructure height is obtained.

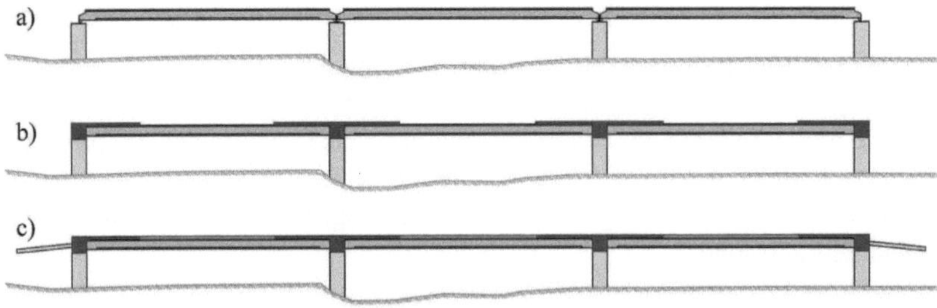

Figure 12:   Superstructure construction phases: laying of the VFT-WIB composite girders in their final position (a), frame node and end sections of the deck concreting (b), deck concreting (c)

3.   The site

Early 2012 the construction of the bridge infrastructure began. The piles, the foundation slabs and the elevations were built.

An indirect foundation made of bored piles, with the concrete class C25/30 was adopted, suitable for integral bridges. One pile beneath each foundation plate includes 3 tubes for the sonic tests ("Fig. 13"). The piles are 8 m, 10 m and 12 m long according to the geological necessities.

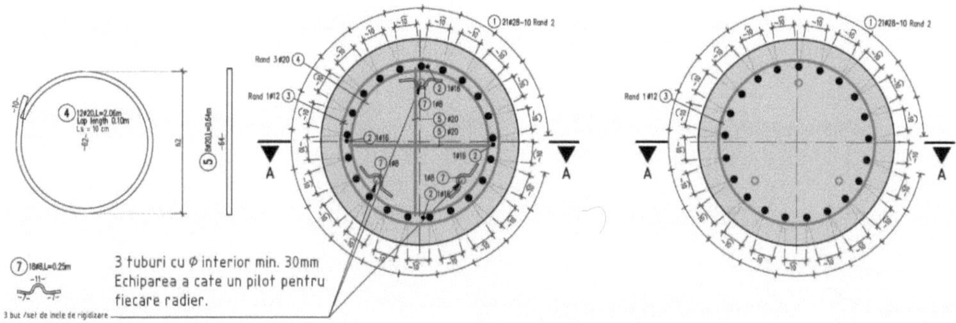

Figure 13:   Pile sections – reinforcement and tubes for the sonic test

In spring 2012 the concrete foundations of class C30/37 were poured. The outgoing reinforcement ensures the connection to the infrastructure elevations ("Fig. 14").

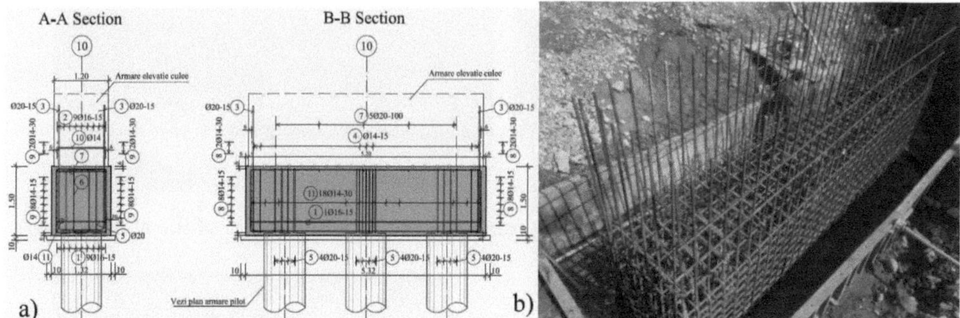

Figure 14:   Foundation slab reinforcement plan (a), Site photo of the foundation slab reinforcement (b)

As a next step the elevations, abutments and piers were built ("Fig. 15", "Fig. 16", "Fig. 17"). To ensure the connection to the superstructure, thus underlining the integral character of the structure, reinforcement bars reach out from the concrete joints.

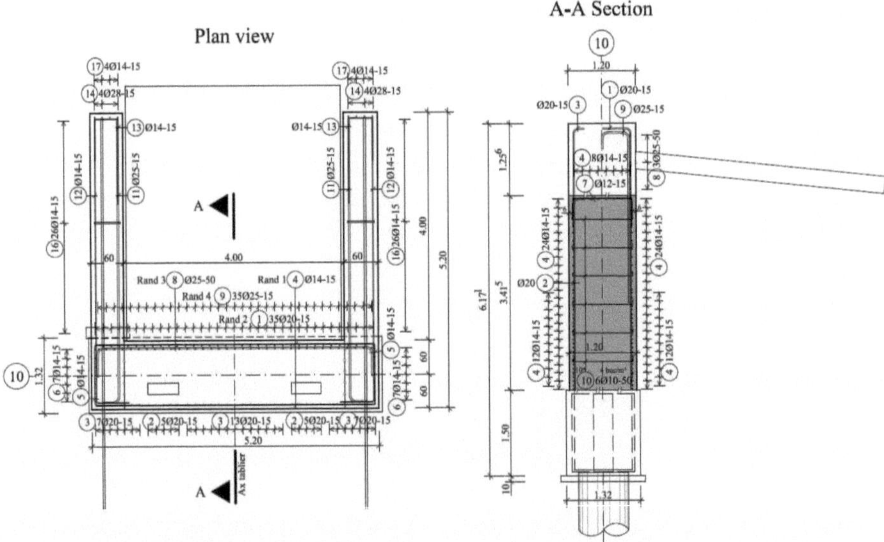

Figure 15:   Abutment reinforcement plan

Figure 16:   Site photo of the abutment reinforcement (a), Site photo of the abutment (b)

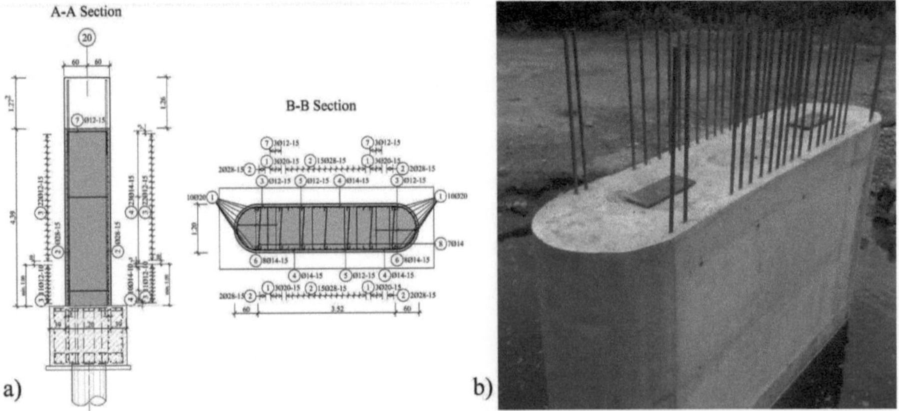

Figure 17:   Pier reinforcement plan (a), Site photo of the pier (b)

The constructive details for the superstructure were designed. These include the formwork and reinforcement of the precast composite girders, the formwork and reinforcement from the cast-in-place concrete deck and the frame nodes reinforcement. The project also includes plans for auxiliary works. For example the river bank protection in area of the bridge was designed using gabions with terramesh modules. Also, the steel parapet was chosen according to the traffic type and volume.

4.   Conclusions

The mono-WIB is an economical solution, as it requires low steel consumption and the prefabrication in the factory is facile. The steel profiles are obtained from regular rolled profiles without significant material loss, and their use is efficient acting as external reinforcement. No bearings and no expansion joints are provided, assuring simultaneously facile maintenance and driving comfort.

Using new technologies for the renewal process fallowing advantages are obtained:

- Robustness of the structure.
- Very slender and aesthetic bridges due to the optimal combination of high tensile strength of structural steel and the high compressive strength of concrete; costs are also minimized.
- High durability of normal reinforced concrete decks due to restrictive crack width limitation.
- Due to the low dead weight of the composite bridges' deck, composite bridges have advantages with regard to the foundation and settlements of supports.
- In comparison with steel bridges, composite highway bridges have a better behavior with regard to deck freezing in winter.
- Due to innovative methods the erection time is very short.
- Reduced environmental impact in comparison with other bridge types.
- Facile and cheap maintenance.

Important delays (even if at that time the bridge in Frânceşti was an absolute premiere in Romania), due to missing financial possibilities postponed the realization of the bridge in Frânceşti. Than the Romanian ECOBRIDGE team took the decision to analyze a new structure situated on the A1 motorway, part of the Orăştie – Sibiu motorway sector, lot 1 at km 8+755, which fulfills all conditions of the initial ECOBRIDGE project. The construction of the bridge at Frânceşti will hopefully soon be resumed, than the local community will safely travel.

References

[1]   RFCS research project, "ECOBRIDGE: Demonstration of economical bridge solutions based on innovative composite dowels and integrated abutment, Grant no. RFSP-CT-2010-00024", 2010

[2]   Pak, D. et al., "Economic and Durable Design of Composite Bridges with Integral Abutments – Final Report RFSR-CT-2005-00041", Brussels, RFCS publications, European Commission, 2009

[3]   Seidl, G, et al., "Prefabricated enduring composite beams based on innovative shear transmission – Final Report RFSR-CT-2006-00030", s.l.: European Commission, 2011.

[4]   Seidl, G, et al., "Economic composite constructions for bridges: construction methods implementing composite dowel strips", Timişoara, Springer Vieweg, ISBN 978-3-658-03713-0, 2013

[5]   Seidl, G, et al., "Composite Dowels in Bridges – Design Guide", D 767, ISBN 978-3-942541-19-9, 2012

Innovative composite overpasses on
the Romanian A1 motorway

Edward Petzek[1], Luiza Toma[2], Elena Meteş[3] & Radu Băncilă[4]

Keywords: VFT® prefabricated composite beam, PreCoBeam, composite dowel, innovative solution, integral bridge, in-situ test.

Abstract: Bridges are of essential importance to the infrastructure; it is a fact that, in the last decades, composite bridges became a well-liked solution in many European countries as a cost-effective and aesthetic alternative to concrete bridges. Their competitiveness depends on several circumstances such as site conditions, local costs of material and staff and the contractor's experience. Beside the classical solution, the new ones with efficient design, which follows the present tendency in bridges consisting in simplifying the structure as much as possible, consolidate the market position of the steel construction and steel producing industry. In the background of Europe's transportation system renewal process, Romania is also passing through a development period of its transportation infrastructure. This is an opportunity for a country to develop efficient, economical and also modern projects, targeting towards obtaining sustainable systems in the end. Continuous shear connection using a cut steel strip is a solution for composite beams. The present paper presents a quick overview of the technical solution chosen for two of the structures from the Orăştie – Sibiu motorway section – overpasses which use in the cross section prefabricated composite girders, applying composite dowels strips and VFT® technology. To prove the accuracy of the model, the bridge was submitted to several in-situ tests, which proved to be a success.

1. Introduction

An important factor within Europe's development process is the transportation system. The completion of the pan-European corridors is also based on Romania's motorway and railway system. Therefore investments in Romania's transportation infrastructure are made.

Many of the new investments in the transportation system are assigned by tender projects in form of „*design & build*" and in this way joint ventures between execution companies

---

1   Assoc.Prof.Dr.Ing., Politehnica University of Timişoara and SSF-RO Ltd. Spl. T. Vladimirescu 12/6, Timişoara, Romania, epetzek@ssf.ro

2   Drd.Ing. Politehnica University of Timşoara and SSF-RO Ltd. Spl. T. Vladimirescu 12/6, Timişoara, Romania, ltoma@ssf.ro

3   Drd.Ing. Politehnica University of Timşoara and SSF-RO Ltd. Spl. T. Vladimirescu 12/6, Timişoara, Romania, emetes@ssf.ro

4   Prof.Dr.Ing., Politehnica University of Timşoara and SSF-RO Ltd. Spl. T. Vladimirescu 12/6, Timişoara, Romania, radu.bancila@ct.upt.ro

and structural engineering offices are given the possibility to build whole road sectors in an economic advantageous manner. This assignment method permits the newly developed, innovative and economical solutions to be used in Romania too.

Therefore a series of efficient solutions of integral bridges made from pre-stressed concrete as well as using steel-concrete composite solutions were implemented in the Romanian motorway system, through the first lot of the Orăştie – Sibiu motorway sector. This project can be characterized by important aspects as: reduced costs, simple and rapid erection – modularity of the system, durability and robustness of the structure, maintenance costs decrease and harmonious aesthetic appearance.

The Orăştie – Sibiu motorway section, with a length of approximate 82 km, crosses the counties Hunedoara, Alba and Sibiu and is part of the IVth pan-European corridor. The route permits design speeds up to 120 km/h, having a minimum radius of the concave fillets of 7000 m, a minimum radius of the convex fillets of 18000 m, a maximum declivity of 2,4% and a minimum declivity of 0,2%. The bridges service lifetime shall be of 120 years with a proper maintenance in accordance with the Romanian standards.

The section was divided in four lots, the assignment procedure being of the type „*design & build*". The first lot, "Planning and building execution of highway Orăştie – Sibiu, contract section 1", near Orăştie and the Mureş River and its confluents, has a length of 24,110 km ("Fig. 1"). It includes 27 bridges as follows:

- 7 motorway bridges, of which 4 over water courses and 3 over other road transport systems (railways, national, county or agriculture roads),
- 7 motorway overpasses for national, county or agriculture roads,
- 13 box bridges, with spans greater than 5,0 m.

All structures are being designed as bridges with integral abutments, except the viaduct from km 1+240, with a total length of 240 m, which is a semi-integral structure. As a consequence, only two pieces of expansion joint equipment and eight bearings were used for the whole motorway lot. This motorway lot can be considered a European premiere, having only integral or semi-integral bridge structures.

From the constructive solutions adopted for this motorway lot we can mention those using in-situ concrete with precast elements for the small span structures (till 12,0m), pre-stressed girder introduced in integral systems (with 36 m girder length), the semi-integral viaduct (with the total length of 240 m) and integral bridges in composite solution – using the VFT® technology (skew bridges using 39 m length composite beams → PreCoBeam).

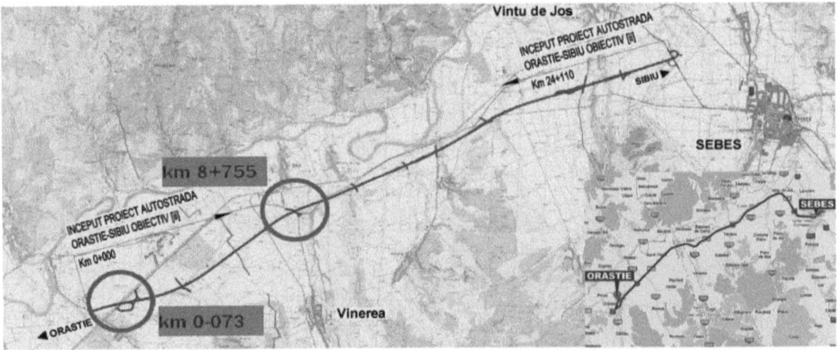

Figure 1:   The motorway route – plan view. The VFT® composite bridges.

Over the last four years, VFT- and VFT-WIB bridges solutions have been applied in Romania. The following structures can be mentioned: the bridge over the Mânărău channel on the national road DN79A or the road bridge over Bistriţa river, country Vâlcea, with a total length of 63,30 m which is under construction, two VFT® overpasses on the A1 motorway, section Orăştie – Sibiu lot 1, one for the county road DJ705 and the second one for the national road DN7.

From the first introduction of the prefabricated composite beams VFT® (Verbundfertigteil – Träger) in the year 1998, through a sustained and systematic research activity several steps forward were made regarding the development of efficient and innovative solutions for composite bridges by creating an efficient connection between the steel and the concrete structure [1]. The research ideas were a success and proved a very good quality and efficiency of the constructive parts (the main girders) which are entirely made in the workshop. The efficiency consists in the elimination of the classical Nelson type connectors and implicitly of the upper flange of the steel section, and creating a new type of *composite dowels* by cutting the steel web according to a predetermined contour ("Fig. 2").

VFT® classic solution   The idea and the new VFT® generation

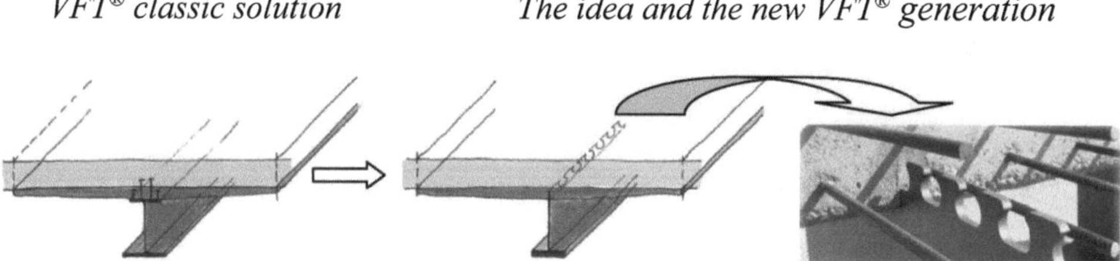

Figure 2:   VFT® solution for composite bridges [1]

2.   Technical details of the applied solutions

In case of the new VFT® the connection between steel and concrete is made through composite dowels and the upper steel flange is needed no more. If the elimination of the upper flanges of the steel girders and the studs are considered, namely material consumption and welding workmanship, the efficiency of the designed solution can be observed. The solution represents an inspired alternative to the classical concrete bridges – particularly to those with pre-stressed girders on the cross section. Using the high degree of prefabrication, the VFT® girders reduce the possibility of unexpected situations on site and offer execution simplicity and by default obtaining lower costs. The reduced weight of the prefabricated composite beams offers advantages both in the transportation as well as in the manipulation efforts and contributes to the success of modern composite bridges.

Both passages have only one span and are oblique (approximately 70°), integral structures and they use four steel-concrete composite beams approximately 39 m long ("Fig. 3").

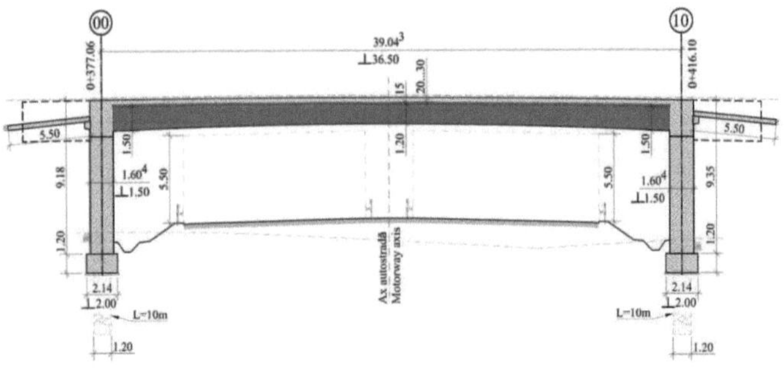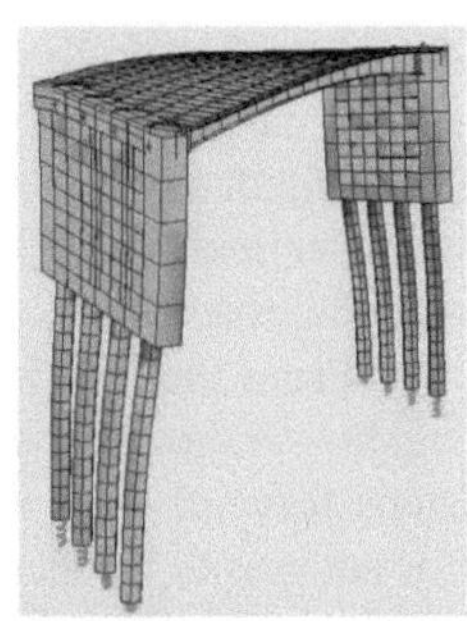

Figure 3:   a) Left: longitudinal section, b) Right: design model

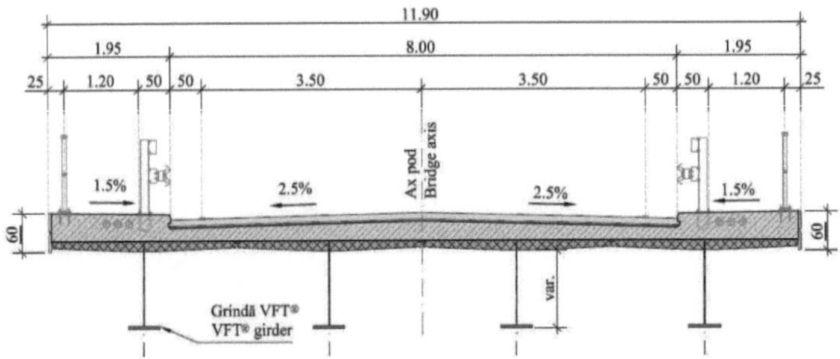

Figure 4:   General cross section [2]

The infrastructure is composed of four drilled piles for every axis – each being 10 m long, continuing a foundation slab – 2,0 m wide and 1,20 m high – and one reinforced concrete straight wall of 1,50 m depth and approximately 10,0 m height ("Fig. 3"). Whilst the infrastructures are made, close to the final bridge position, four prefabricated steel girders are brought from the steel plant and are fixed on temporary concrete supports, designed to take into account the precambering of the girder. On those temporary platforms the upper reinforced concrete flange of the precast composite girders begins to take shape with its variable thickness 10 cm ÷ 15 cm ("Fig. 5", "Fig. 6"). The concrete flange serves as a pressure zone, horizontal bracing during erection and formwork for the in-situ slab.

Then, the finished precast composite beams – the VFT® girders are aligned on the infrastructure in the final position. During the design stages and also during execution, special attention was paid to the reinforcement in the frame corners ("Fig. 7"). In accordance with the designed technology, the pouring of the superstructure plate was made in two steps: first on the frame corners areas and only afterwards the works were continued in the middle part of the superstructure. The superstructure plate is 20 cm ÷ 30 cm thick.

The last remaining work was the proper equipment of the structure: the waterproofing system, the pedestrian parapets, the guardrail protection, the translation slabs, earth filling with the corresponding degree of compaction and adequate materials, road structure ("Fig. 8).

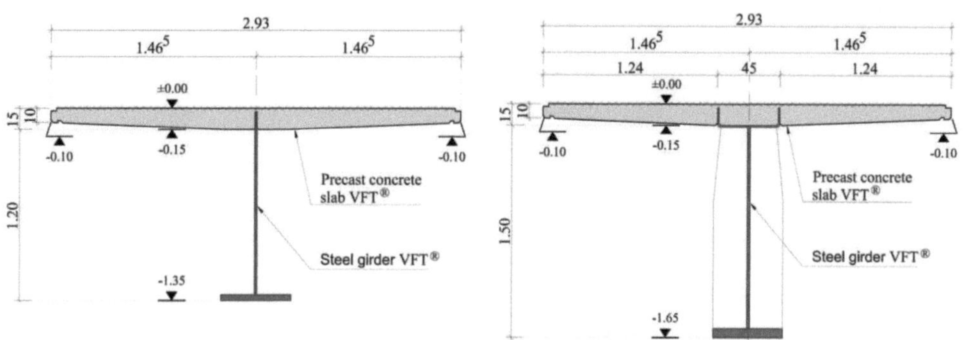

Figure 5:   The formwork cross section of the Prefabricated Composite Girder VFT®

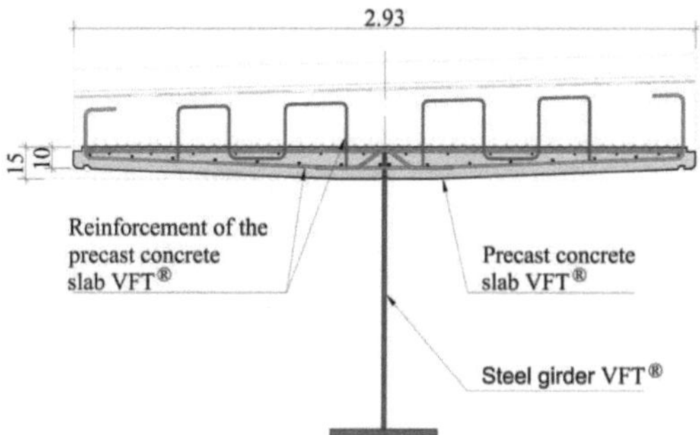

Figure 6:   Reinforced cross section of the Prefabricated Composite Girder VFT®

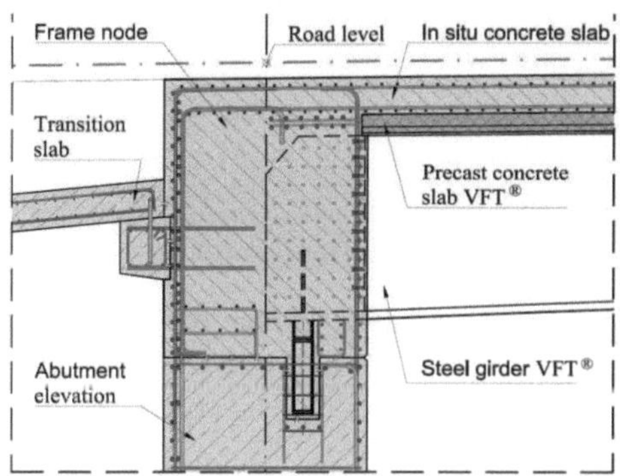

Figure 7:   Frame node reinforcement detail

Figure 8:   The final structure and its equipment.

The welded steel girders from the prefabricated composite beams are made of S355 J2+N and have variable height – 1200 mm in the field and 1500 mm at the supports. Two types of section were introduced: the middle part of the steel girder is composed only of a bottom flange and a web with steel dowel cuttings on top and on the end zones, of each 6 meters, a slender U-shaped upper flange having a thickness of 12 mm was additionally introduced ("Fig. 9"). The steel plates of the upper flange were cold bended and have the role to increase the number of the steel dowels in order to transfer the shear effect.

The composite dowels require a cutting geometry on the upper part of the steel ("Fig. 9", "Fig. 10") using a special form adapted to the requirements given by fatigue, reinforcement and concrete.

Figure 9:   The steel girder type sections; Steel girders in the workshop; the cutting line geometry of the composite dowels.

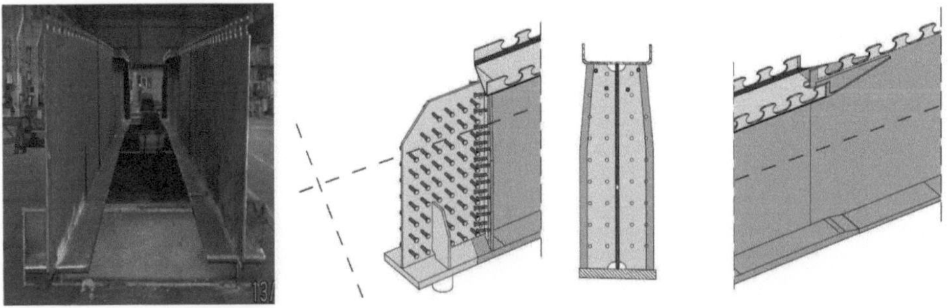

Figure 10:   10 The steel girder type sections; the end steel girders detail and the transition detail between the two types of section.

A very slender and aesthetic aspect was obtained for these overpasses and the solution allowed considerable material savings. Practically, for both bridges a very good steel consumption of 130 kg/m² was obtained.

Figure 11: Finished overpasses a) Structure 01 – km 0-073; b) Structure 11 – km 8+755

3. Technical details of the load test project

A load test was carried out for one of the twin structures – the overpass at km 8+775. The test program was conceived according to the Romanian standard [5]. The test purpose was to obtain information about the structure behaviour subjected to present load situations, to assess the viability of the initial design project.

According to the Romanian standard, the analysis of the following issues must be taken into account within a testing program:

- the resistance and the stability of the structure;
- the superstructure's elastic behaviour;
- the state of deformations and displacements in characteristic points and sections;
- functional behaviour;
- the overlapping of in-situ results with those obtained in the calculation design stage.

According to the national regulations it is compulsory to test all the new bridges which present novelties from the point of view of the used material, the method of calculation and execution technologies. The tests will be performed under static and dynamic actions.

The entire testing process contains some well-established conditions, as follows:

- Several consecutive cycles of load schemas.
- In the static load case, the truck convoys entered and exited the bridge with the maximum speed of 5 km/h. They stopped in the position specified in the load scheme;
- In case of dynamic loads, the convoys of motor vehicles circulate on the bridge with a constant speed.
- During the test the circulation on the bridge is closed.
- To increase the impact on the tested structure, it is recommended to create artificial unevenness of the way, by placing on the road wooden planks with the thickness of 4 cm, width of 30 cm and length of 300 cm. The edges from the superior part of the wooden planks were flattened at 45°. Their positions on the bridge were established by the testing design.
- The testing actions included following information on the vehicles: the weight on the axle, position of the axles and of the front wheels in relation to the elements of the resistance structure of the bridge.
- For the dynamic testing of the bridge the speed of movement of the convoy must be controlled and their dimensions, position and weight must be also known.
- A visual observation of the bridge was also made.

According to the Romanian standard [5], the load test vehicles convoy must be similar to the A30 theoretical convoy, available on the market. Thus MAN TGA 41.480 trucks, type BB – with four-axle were used. To keep a realistic situation, the trucks were loaded to the maximum weight of 40 t, closer to the present traffic situation on this national road ("Fig. 12", "Fig. 13"). A number of two static load schemas and four dynamic load schemas were provided, a total of maximum four MAN TGA trucks per schema being used.

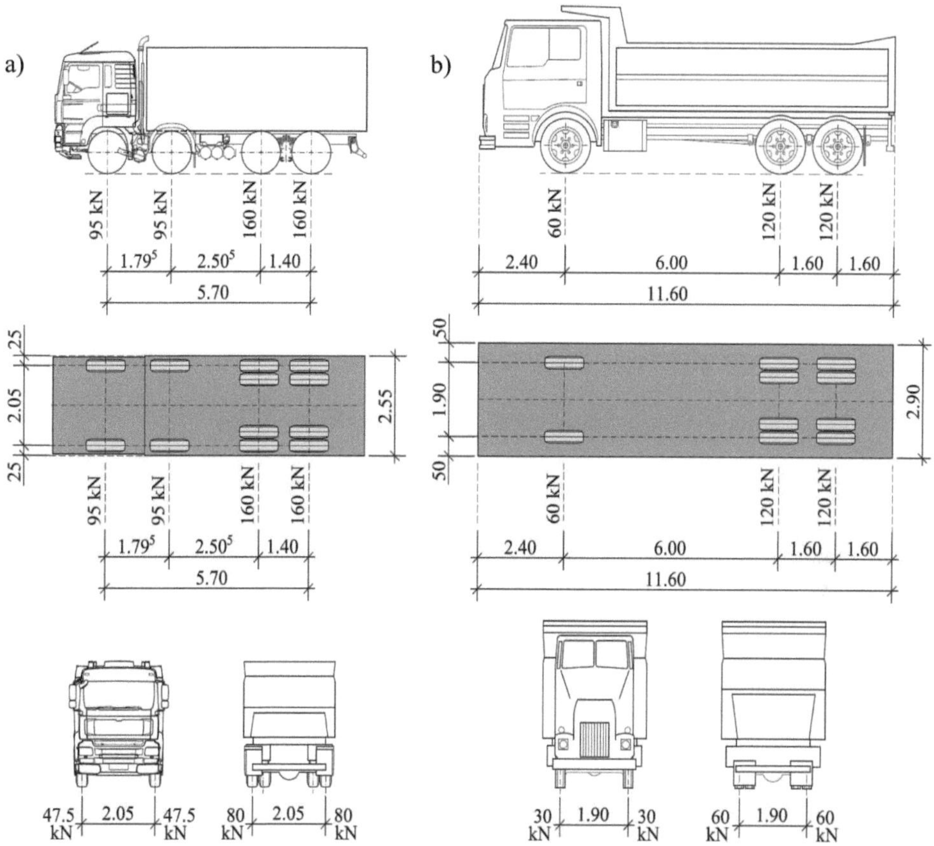

Figure 12: a) MAN TGA convoy – lateral view, plan view, front / back view;
b) A30 convoy – lateral view, plan view, front / back view.

Figure 13: MAN TGA convoy – front view, lateral view; dynamic load test MAN TGA vehicle.

In order to obtain relevant results – the maximum stress and displacements in the structure – six different load schemas were established. The ratio between the measured values and the values obtained in the FEM analysis of the structure is defined in "Eq. 1".

$$E_{stat} = \frac{S_{measured}}{S_{design}}$$
(1)

Where:

$S_{measured}$ – the designed efforts, which are verified according to the method of limit states;

S_{design} – the stress obtained from the testing loads considered as statically applied.

For the dynamic test the dynamic coefficient was determined "Eq. 2".

$$\Psi_{mas} = \frac{\varepsilon_{din}}{\varepsilon_{stat}}$$
(2)

Where:

ε_{din} – maximum specific deformation measured in a point, at the passing of the dynamic convoy with a certain speed,

ε_{stat} – specific deformation measured in the same point where it has been measured ε_{din}, from the loading with the same convoy applied statically in the position in which ε_{din} was produced.

Before starting the effective procedures, the following preparatory works were realized:

- The entire team of participants: authorities, engineers, technicians are present on the site.
- The load tests, the strain gauges, the cables and all the necessary equipment for the test were prepared in advance.
- All the measurement devices were verified a few days before.
- Prior to the installation of the strain gauges, the surfaces from the steel beams were prepared according to the usual procedure.
- One day before the load test, the strain gauges on the structure, indicated by the drawing, were installed in accordance with the test project ("Fig. 14"). The strain measurements were made in two different sections, for each steel beam: the first section is located near the frame corner, where two active strains on the flanges and one passive strain on the neutral axis ("Fig. 15a", "Fig. 16a") are disposed and the second one is the middle cross section with one pair of strains – the active one on the bottom flange and the passive one in the neutral axis ("Fig. 15b" , "Fig. 16b").
- The vehicles' real weight on the axle was verified and then filled out in the prepared forms from the project.
- The axles and wheels positions were marked properly on the bridge according to the design drawings.
- The equipment provided for the deformations measurements was installed and the required topographic devices were prepared;

- The weather conditions on the test load day had to be taken into consideration by the engineers (temperature, wind and precipitations).

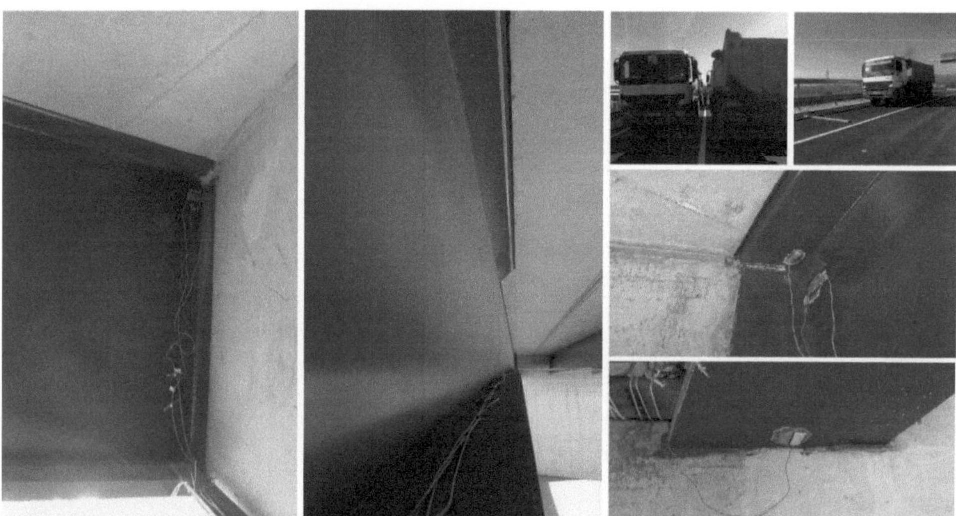

Figure 14: Installed strains on the section of the bridge in the in-situ test day

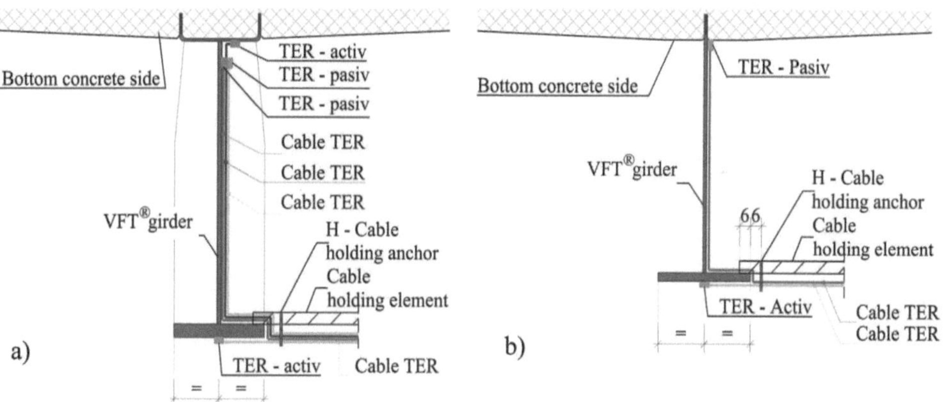

Figure 15: Installed strains on the section of the bridge; a) Cross section of the supports area; b) Cross section of the field.

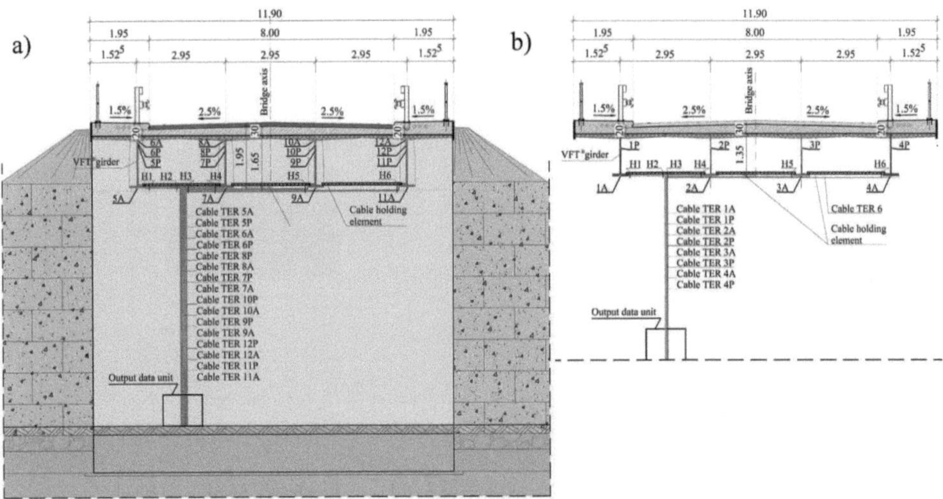

Figure 16:   Installed strains on the section of the bridge; a) Cross section on the supports area; b) Cross section on the field.

The measuring devices were meant to be read in the following steps: prior to the loading, during the load schema and after each discharge. During the reading procedure of all the measurement devices (stresses or displacements) the state of effort of the bridge structure was maintained constant. Before the testing the preliminary readings provided by the standard were performed: twice with approximately an hour ahead, respectively 15 minutes before the beginning of the testing. After the performance of each scheme of loading, as well as after each unloading, the measurement devices are read many times. The first reading is effectuated immediately after the disposal of the testing actions on the bridge, respectively after unloading, than readings at equal intervals in time (of at least 5 minutes) are performed until the stabilization of the measured values, namely until the increasing of the values measured between two consecutive readings is under 15%.

The principle of the strain gauge chain used for the tests is presented in "Fig. 17". On the structure on which the loads $F_1 \ldots F_n$ act, the strain gauge is applied (glued) (TER – transducer with electrical resistance) and connected through the measuring cable with several conductors (1) to the amplifier. The data acquisition is made with the help of the laptop connected to the amplifier through the interface (2).

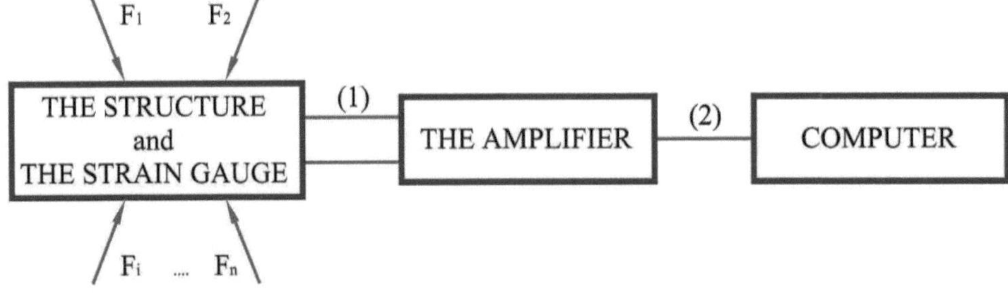

Figure 17:   The principle of the strain gauge chain used for the probes.

Consequently, as the static schema of the bridge is a frame and the expansion joints were eliminated, it appeared as necessary and important to check the longitudinal displacements of the entire assembly. These measurements were provided by placing a displacement device – type W.A.1 on the back of one of the frame corners, before the disposal of the transition slab, the earth filling and the road structure. Using an extension, the W.A.1 cable was brought out to the outside surface of the sidewalk of the bridge, where with the help of a reading device, the results could be recorded and read off any time it was wanted ("Fig. 18"). The measurements must be made at least once per season to obtain the displacement between extreme temperatures, especially winter-summer differences, as well as day-night differences. The monitoring has to be done at least one year, in order to obtain in the end a graphic with the movement differences of the structure.

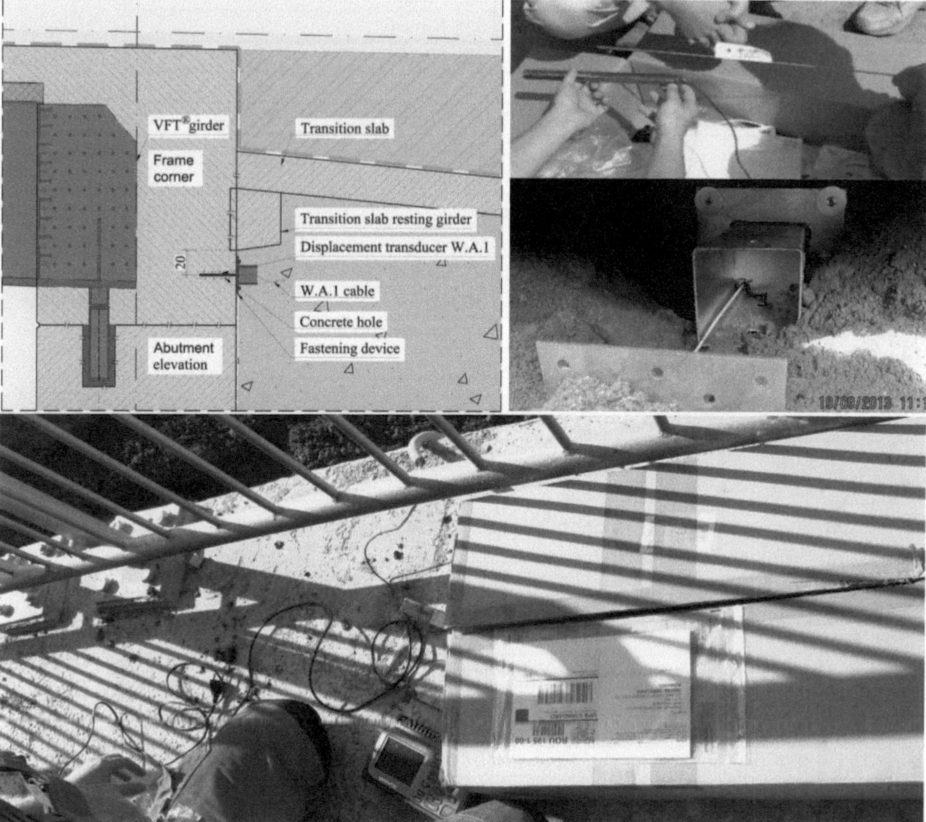

Figure 18: Displacement transducer device

4. Designed aspects of the assembly

For the static calculation a design platform based on the FEM analysis of the structure was used, obtaining an accurate simulation of its behaviour in comparison with real forces, vibration, heat, possible vertical displacements and other physical effects. Using this kind of computerized method the initial design was validated and optimized, before releasing the project to the constructor. The bridge components, the VFT girders, the concrete slab, the piles are defined as linear elements, except the infrastructures which are represented as block elements, all being in concordance with the dimensions from the project and having real boundaries conditions. The piles were defined as circular cross sections, d = 1,20 m, affected by springs, the pile bedding constants being in correlation with the provided geotechnical investigations. The materials used for the entire structure are according to the Eurocode prescriptions ("Tab. 1").

Table 1:   The structure material list

Materials				
S355 J2	$t_w <= 40$ mm	f_{yk}	355 N/mm²	main beams
		f_{yd}	355 N/mm²	
	$t_w > 40$ mm	f_{yk}	335 N/mm²	
		f_{yd}	335 N/mm²	
	E_A		210000 N/mm²	
	Standards		SR EN 1990, SR EN 1993, EN 10025-2	
C25/30 XA1, XC2 S4 W/C=0,55 D_{max} = 16		f_{ck}	25 N/mm²	piles
		f_{cd}	14 N/mm²	
	E_{CM}		31000 N/mm²	
	E_A / E_{CM}		—	
	Standards		SR EN 1990, SR EN 1992, SR EN 1997	
C30/37 XA1, XF3, XC2 S4 W/C=0,50 D_{max} = 16		f_{ck}	30 N/mm²	foundation slab
		f_{cd}	17 N/mm²	
	E_{CM}		32000 N/mm²	
	E_A / E_{CM}		—	
	Standards		SR EN 1990, SR EN 1992, SR EN 1997	
C30/37 XD1, XF2, XC4 S3 W/C=0,50 D_{max} = 16		f_{ck}	30 N/mm²	abutments
		f_{cd}	17 N/mm²	
	E_{CM}		32000 N/mm²	
	E_A / E_{CM}		—	
	Standards		SR EN 1990, SR EN 1992, SR EN 1997, INTAB+	
C35/45 XD3, XF2, XC4 S4 W/C=0,45 D_{max} = 16		f_{ck}	35 N/mm²	in-situ concrete
		f_{cd}	19,8 N/mm²	
	E_{CM}		34000 N/mm²	
	E_A / E_{CM}		—	
	Standards		SR EN 1990, SR EN 1992	
C45/55 XD3, XF2, XC4 S4 W/C=0,45 D_{max} = 16		f_{ck}	45 N/mm²	precast slab
		f_{cd}	25,5 N/mm²	
	E_{CM}		36000 N/mm²	
	E_A / E_{CM}		5,83	
	Standards		SR EN 1990, SR EN 1992	
Bst 500 SC		f_{yk}	500 N/mm²	all elements
		f_{yd}	435 N/mm²	
	Standards		SR EN 1990, SR EN 1992	

Another important aspect of the modelling concept was the design using construction stages, in this way the behaviour of the structure was studied and improved at every step provided by the technology. The initial elements of the bridge were considered to act as individual parts and they were included step by step in the entire assembly, initially a simple supported girder boundary condition for the VFT girders was provided and then the frame effect in the calculation in the right stage was captured ("Fig. 19"- "Fig. 21").

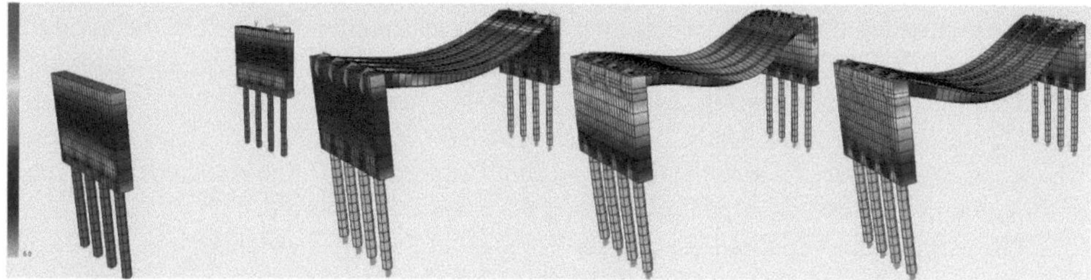

Figure 19:   Fig. 19 Construction stage manager steps.

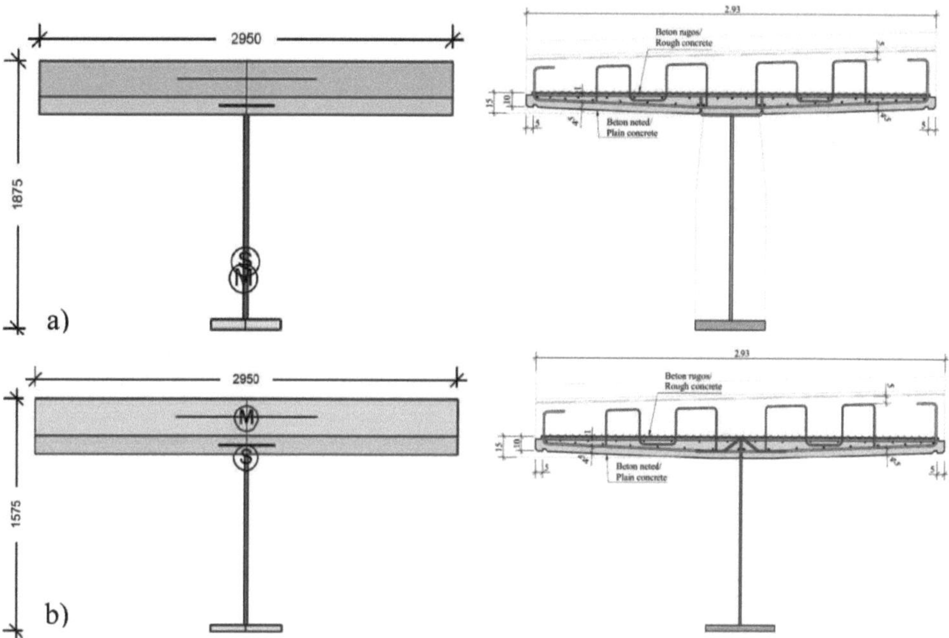

Figure 20:   a) Cross section of the supports area; b) Cross section of the field.

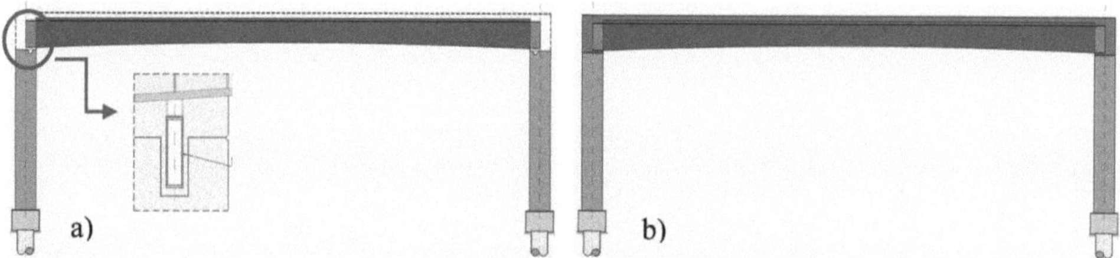

Figure 21:   Construction stags: a) Simple supported girder; b) Frame effect

On the structure thus defined two loads system were considered: the Eurocode tandem system and also the loads similar to those resulted from the testing convoys – the MAN TGA vehicles were introduced ("Tab. 2", "Fig. 22").

Concerning the static load test schemas two cases – LC1, LC2 have been considered. In the first situation the convoys consisting of four trucks were positioned in the middle of the span, being centred on the cross-section, in the second case the trucks were moved to the outside of the cross-section ("Fig. 23").

For the dynamic test a MAN TGA truck type BB (equivalent to A30) was considered, which was running parallel to the highway axis with a speed of 30 km/h and then with a speed of 50 km/h. In this case, four load schemas – LC1…4 were conceived.

The temperature variation, bearing settlements, all the self – weights including the pedestrian load were modelled on the structure, for the ultimate limit state and as well as for the service limit state.

Table 2:   The load test vehicles weights

The load test vehicles (MAN TGA)	Axle number	Weight / axle [to]	Load / wheel [kN]
1	1	7.030	35.150
	2	7.030	35.150
	3	11.830	59.150
	4	11.830	59.150
	Total weight	37.720	
2	1	7.590	37.95
	2	7.590	37.95
	3	12.780	63.90
	4	12.780	63.90
	Total weight	40.740	
3	1	6.990	34.95
	2	6.990	34.95
	3	11.760	58.80
	4	11.760	58.80
	Total weight	37.500	
4	1	7.000	35.00
	2	7.000	35.00
	3	11.770	58.85
	4	11.770	58.85
	Total weight	37.540	

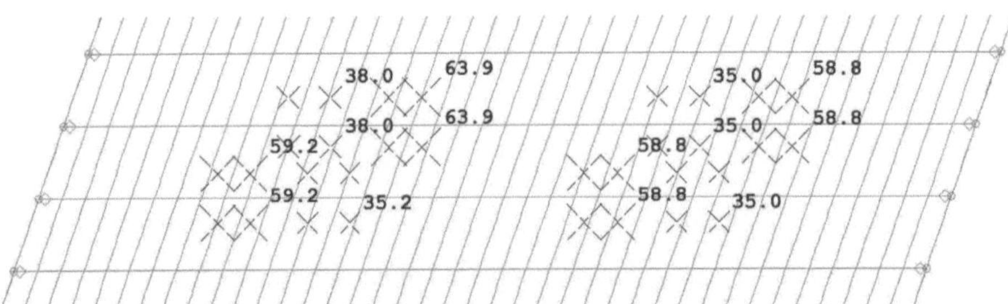

Figure 22:   The load test vehicles designed in the FEM model

In the figure "Fig. 23" all established read points P1 – P12 for the static / dynamic test schemas are represented. The displacements were measured only in the middle of the structure (P1 – P4) with the help of the simple wire plumb principle and a scale, but also with the help of a vertical displacement device ("Fig. 24").

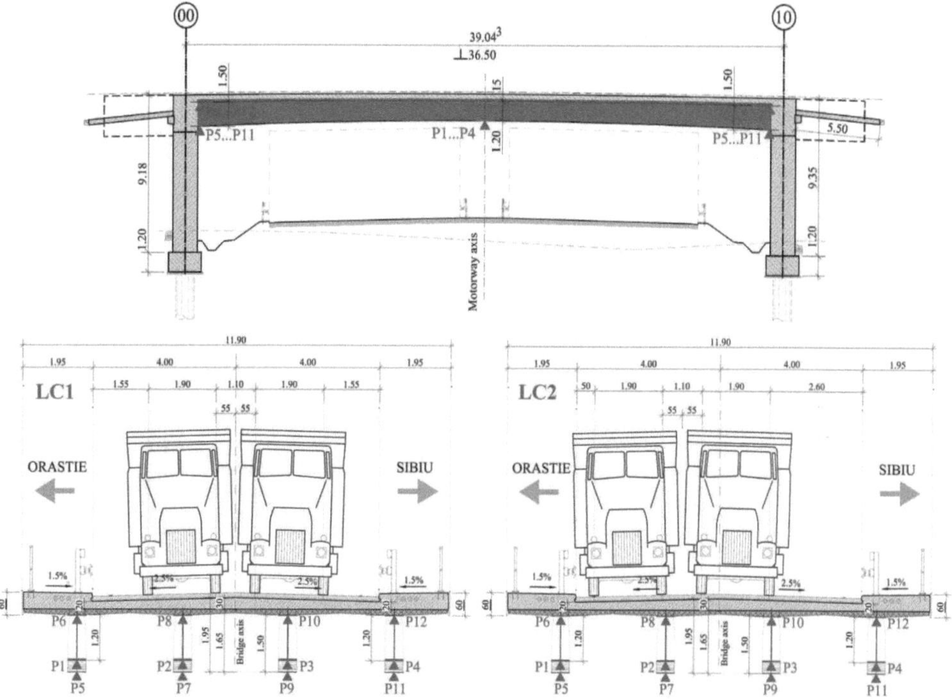

Figure 23:   The read points for the static / dynamic test schema, load schema 1 – LC1 and load schema 2 – LC2.

Figure 24:   Vertical displacement measuring principle

5.   Data processing. Results

The results obtained during the bridge test were interpreted and compared with the model calculation values according to the methodology described by the standards.

The report between the ratio of the measured elastic deformation (f_{el}) and the calculated ones (f_{teor}) ("Fig. 25") were corresponding to the criterion of similitude, the maximum values $E_{f\,stat}$ in the middle of the span, the design characteristic loads being compared to the testing loads.

Table 3:   Maximum vertical displacements for load schema 1 – LC1 and load schema 2 – LC2

Load case	Section item	Recorded values			Vertical displacement		
		unloaded structure	loaded structure	discharge structure	f_{el} [mm]	f_{teor} [mm]	Δ
1	P1	3305	3312	3306	7.00	7.90	1.13
	P2	2718	2726	2719	8.00	9.00	1.13
	P3	2115	2123	2117	8.00	9.00	1.13
	P4	3493	3500	3494	7.00	7.80	1.11
2	P1	3306	3314	3306	8.00	9.00	1.13
	P2	2719	2727	2719	8.00	9.50	1.19
	P3	2117	2124	2118	7.00	8.50	1.21
	P4	3495	3500	3495	5.00	6.00	1.20

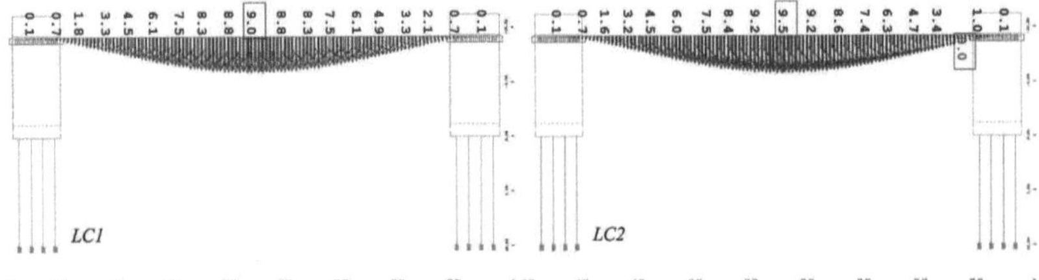

Figure 25:   Vertical displacement diagrams – test load schema 1 – LC1 and load schema 2 – LC2

Table 4:   Results from test load schema 1 – LC1

Section item	$s_{measured}$ N/mm²	s_{design} N/mm²	$E_{f\,stat}$
P1_*Test channel 9*	23.52	23.80	0.99
P2_*Test channel 10*	27.30	27.13	1.01
P3_*Test channel 11*	28.54	27.22	1.05
P4_*Test channel 12*	22.72	23.16	0.98
P6_*Test channel 1*	27.72	28.84	0.96
P8_*Test channel 2*	35.09	38.28	0.92
P10_*Test channel 3*	36.09	38.73	0.93
P12_*Test channel 4*	27.68	26.98	1.03
P5_*Test channel 5*	-19.92	-19.72	1.01
P7_*Test channel 6*	-28.55	-32.70	0.87
P9_*Test channel 7*	-33.06	-32.93	1.00
P11_*Test channel 8*	-20.06	-20.80	0.96

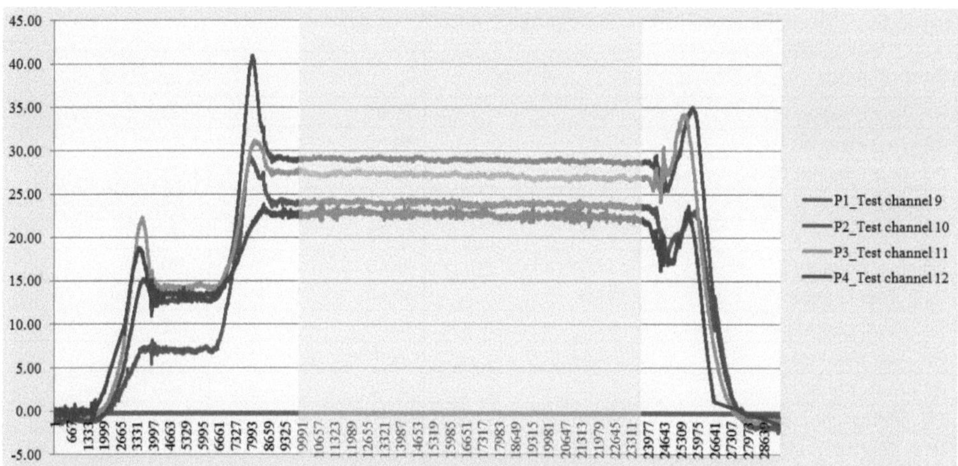

Figure 26: „$s_{measured}$" values from test load schema 1 – LC1

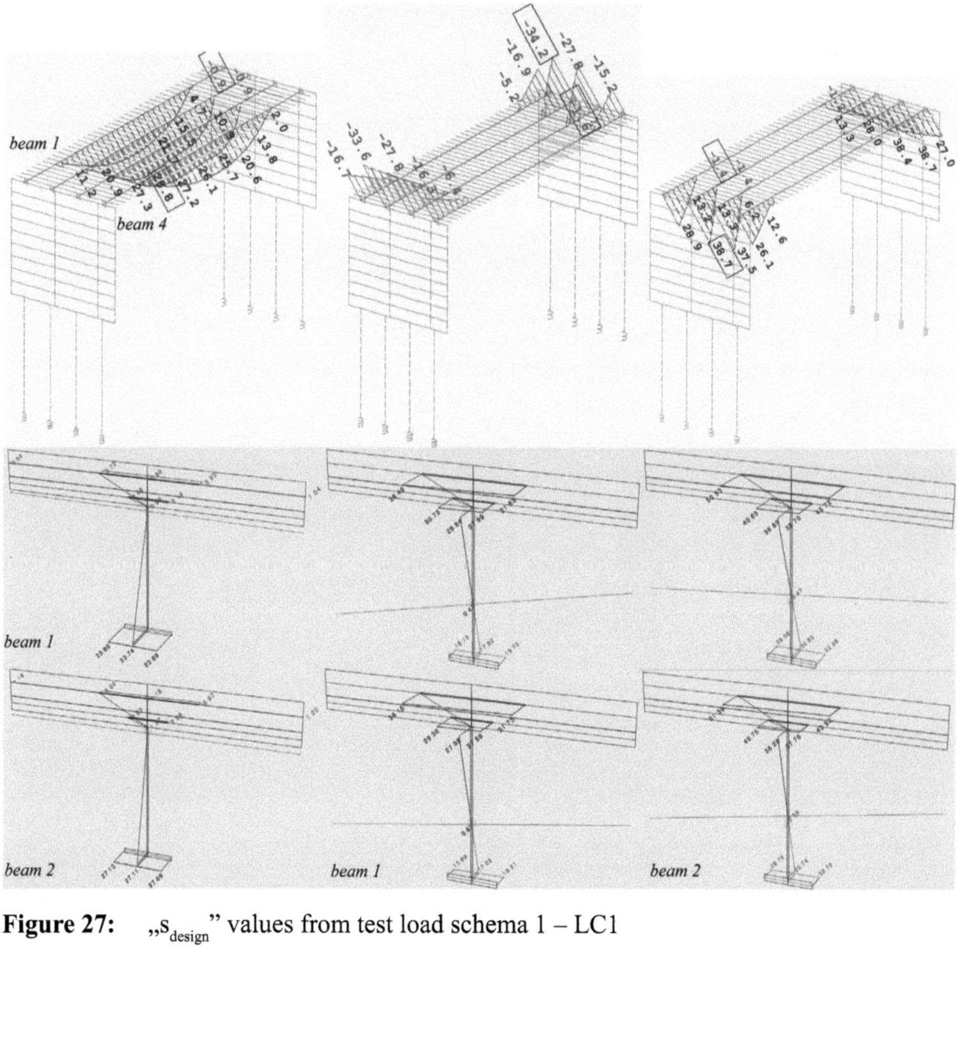

Figure 27: „s_{design}" values from test load schema 1 – LC1

Table 5:   Results from test load schema 2 – LC2

Section item	$s_{measured}$ N/mm²	s_{design} N/mm²	$E_{f\,stat}$
P1_*Test channel 9*	31.50	31.95	0.99
P2_*Test channel 10*	30.45	30.60	1.00
P3_*Test channel 11*	24.15	23.80	1.01
P4_*Test channel 12*	15.75	16.03	0.98
P6_*Test channel 1*	38.51	40.68	0.95
P8_*Test channel 2*	37.83	42.13	0.90
P10_*Test channel 3*	33.96	34.44	0.99
P12_*Test channel 4*	16.07	15.73	1.02
P5_*Test channel 5*	-29.44	-28.75	1.02
P7_*Test channel 6*	-29.85	-36.21	0.82
P9_*Test channel 7*	-26.16	-30.10	0.87
P11_*Test channel 8*	-14.32	-14.84	0.96

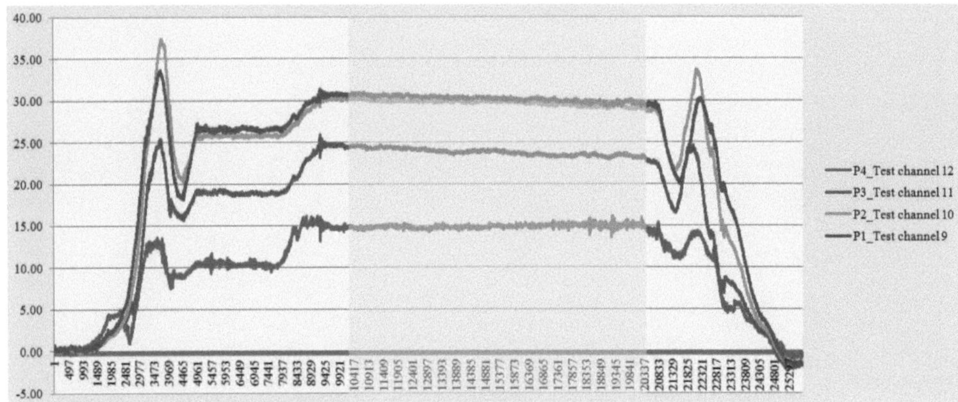

Figure 28:   „$s_{measured}$" values from test load schema 2 – LC2

Figure 29:   Preparatory actions for the load test convoy – LC2

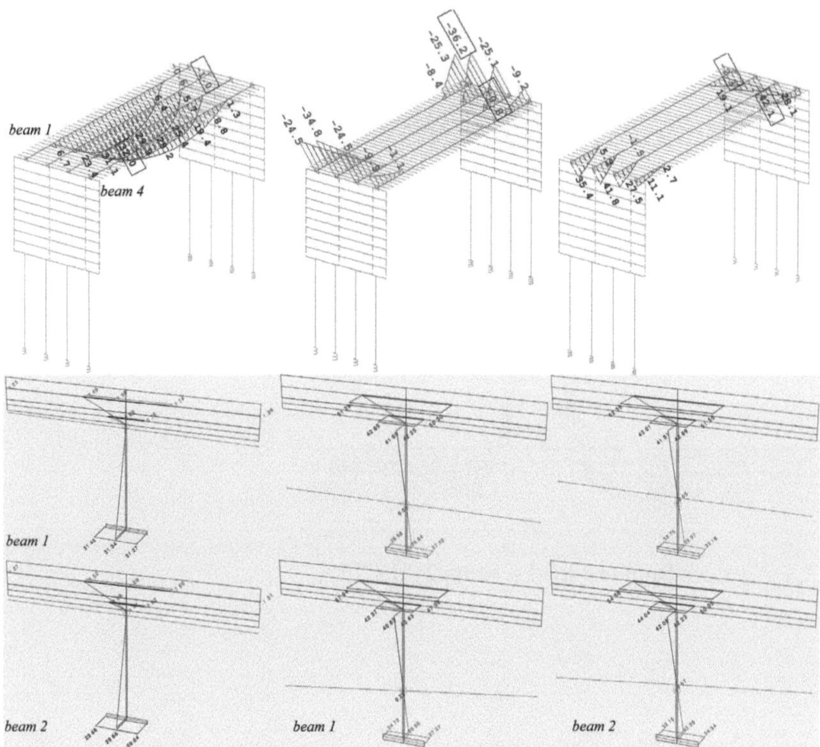

Figure 30: 30 „s_{design}" values from test load schema 2 – LC2

Table 6: Results from dynamic load test schema 2 and 5 – LC2, LC5.

Load case	Section item	s_{din} N/mm²	s_{stat} N/mm²	ψ_{mas}
LC2	P1_Test channel 9	19.50	15.04	1.30
	P2_Test channel 10	23.86	17.33	1.38
	P3_Test channel 11	14.56	10.77	1.35
	P4_Test channel 12	4.70	3.44	1.37
LC5	P1_Test channel 9	20.20	15.04	1.34
	P2_Test channel 10	23.68	17.33	1.37
	P3_Test channel 11	14.60	10.90	1.34
	P4_Test channel 12	4.60	3.44	1.34

Table 7: Results from dynamic load test schema 3 and 4 – LC3, LC4.

Load case	Section item	s_{din} N/mm²	s_{stat} N/mm²	ψ_{mas}
LC3	P1_Test channel 9	21.56	15.76	1.37
	P2_Test channel 10	24.67	17.94	1.38
	P3_Test channel 11	14.47	11.00	1.32
	P4_Test channel 12	4.53	3.50	1.29
LC4	P1_Test channel 9	21.65	15.76	1.37
	P2_Test channel 10	24.79	17.94	1.38
	P3_Test channel 11	14.78	11.00	1.34
	P4_Test channel 12	4.60	3.50	1.31

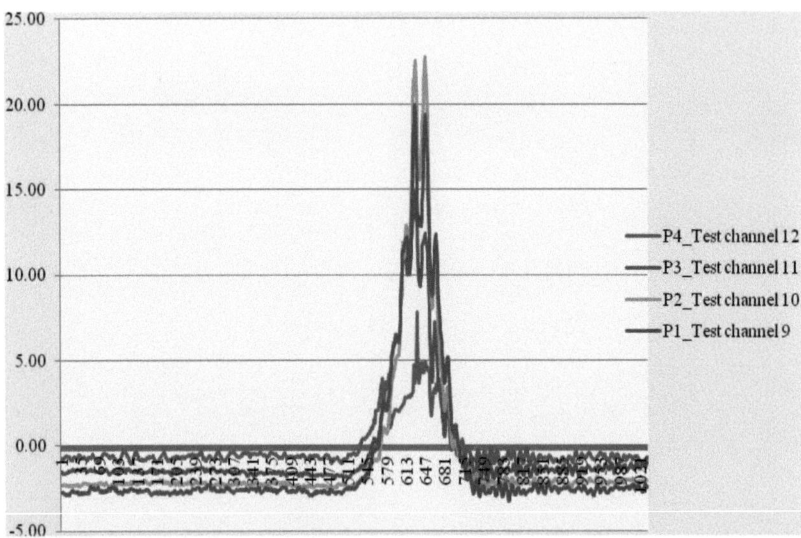

Figure 31: „s_{din}" values from test load schema 2 – LC2

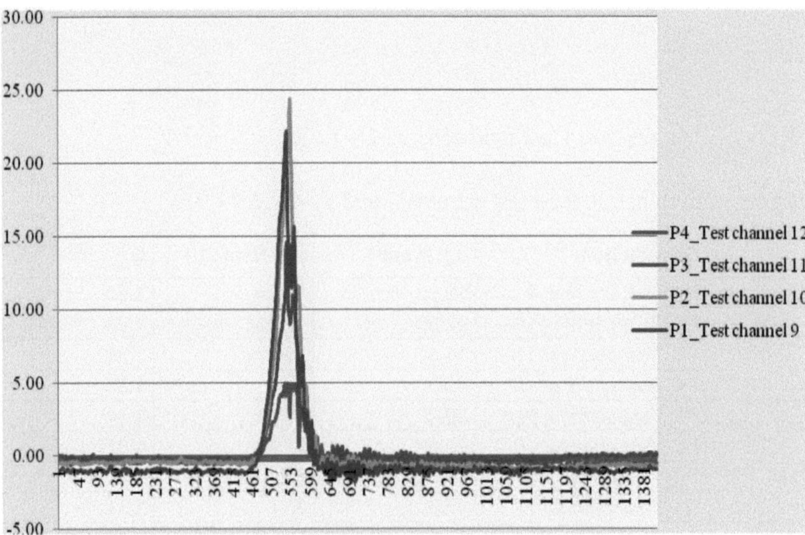

Figure 32: „s_{din}" values from test load schema 3 – LC3

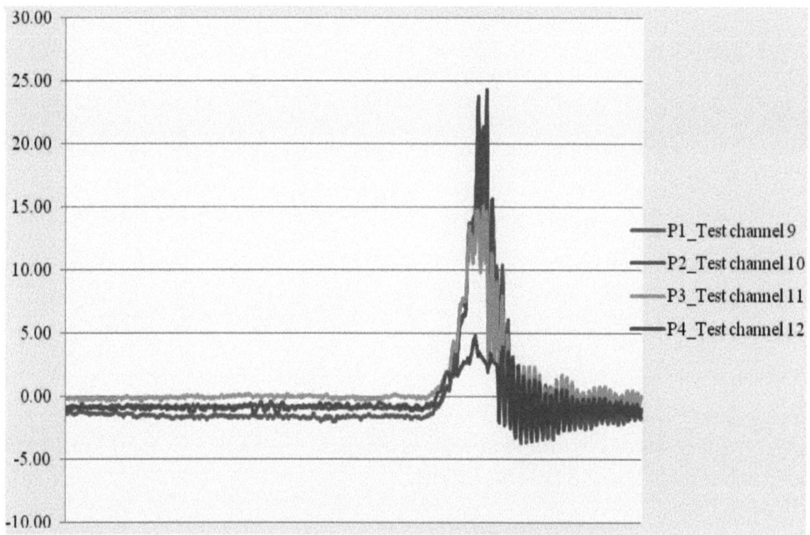

Figure 33: „s$_{din}$" values from test load schema 4 – LC4

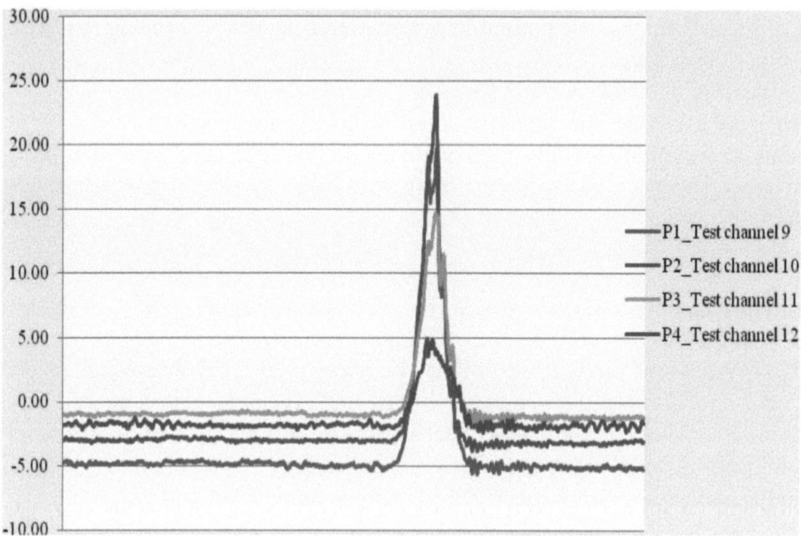

Figure 34: „s$_{din}$" values from test load schema 4 – LC5

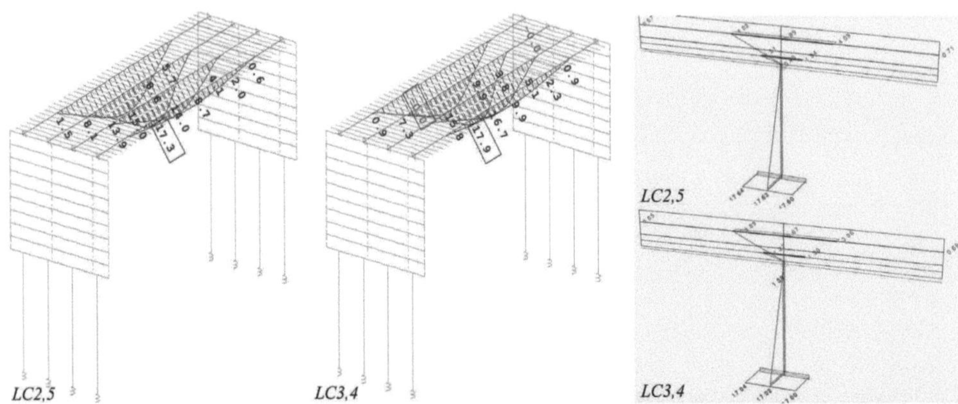

Figure 35:   „s_{stat}" values from dynamic test load schema 3, 4 – LC3,4 (v = 50 km/h) and
load schema 2 – LC2,5 (v = 30 km/h)

As load test conclusions, the structure has had a proper behaviour considering the following aspects:

- the bridge or its elements do not present important deformations, exceeding the elastic ones, or losing of stability;
- at a close inspection no cracks were observed;
- flaws which could affect the functionality of the bridge did not appear;
- displacements were within the limits recommended by the technical documentation;
- the unitary stresses in the areas with maximum stress are close to those resulted from the calculus.

6.   Conclusions

In case of the two overpasses from Orăştie – Sibiu Motorway – lot 1, the theoretical concept – the new VFT® cross section was put in practice for the first time in Romania, as well being a engineering premiere. Combining the theoretical aspects of the new solution with a proper technology a significant material reduction with a minimum steel consumption of 130 kg/m² was obtained; in the end having two slender, robust and economical structures.

7.   Project participants

For both motorway bridges the project participants were:
Client   Compania Naţională de Drumuri şi Autostrăzi din România
Construction company   STRABAG s.r.l
Engineering structures   SSF-RO s.r.l. and SSF Ingenieure AG

References

[1]   Seidl, G., et alt.: Prefabricated enduring composite beams based on innovative shear transmission, Final Report RFSR-CT-2006-00030, s.l.: European Commission, 2011.

[2]   s.c. SSF-RO s.r.l.: Project 11-1024, Bridges design of the Orăştie-Sibiu, lot 1 km 0,000-24,110, Motorway section.

[3]   RFCS research project 2010. ECOBRIDGE: Demonstration of economical bridge solutions based on innovative composite dowels and integrated abutment, Grant no. RFSP-CT-2010-00024.

[4]   Petzek, E., Băncilă R., Schmitt V.: Romanian experience in new efficient solutions for composite bridges, Eurosteel 2011 Budapest, 978-954-724-014-5, 2011.

[5]   Romanian Standard "STAS 12504-86, Validation date: 1986-11-01"

MIX
Papier aus verantwortungsvollen Quellen
Paper from responsible sources
FSC® C105338

If you have any concerns about our products,
you can contact us on
ProductSafety@springernature.com

In case Publisher is established outside the EU,
the EU authorized representative is:
**Springer Nature Customer Service Center GmbH
Europaplatz 3, 69115 Heidelberg, Germany**

Printed by Libri Plureos GmbH
in Hamburg, Germany